GENE GAME

THE NEW LAWS OF EVOLUTION

By

George R. vB. Ennenga

authorHOUSE®

AuthorHouse™
1663 Liberty Drive
Bloomington, IN 47403
www.authorhouse.com
Phone: 1-800-839-8640

First published by AuthorHouse 8/10/2010

ISBN: 978-1-4389-1411-4 (sc)

Printed in the United States of America

This book is printed on acid-free paper.

BLOG: www.genegame.blogspot.com

Electronic Book available

DEDICATION

To the Wonder and Majesty before us,
Around us, within us,
To the Glory Expanding,
And, to our futures.

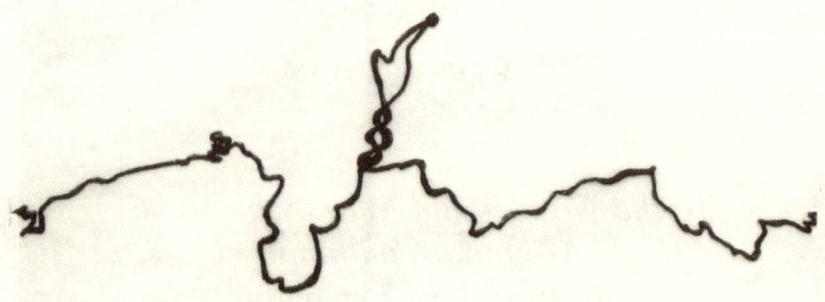

"These facts seemed to me to throw some light on the origin of species- that mystery of mysteries, as it has been called by one of our greatest philosophers."

Charles Darwin, referring to Aristotle

CONTENTS

LIST OF ILLUSTRATIONS

LIST OF ILLUSTRATIONS (CONTINUED)

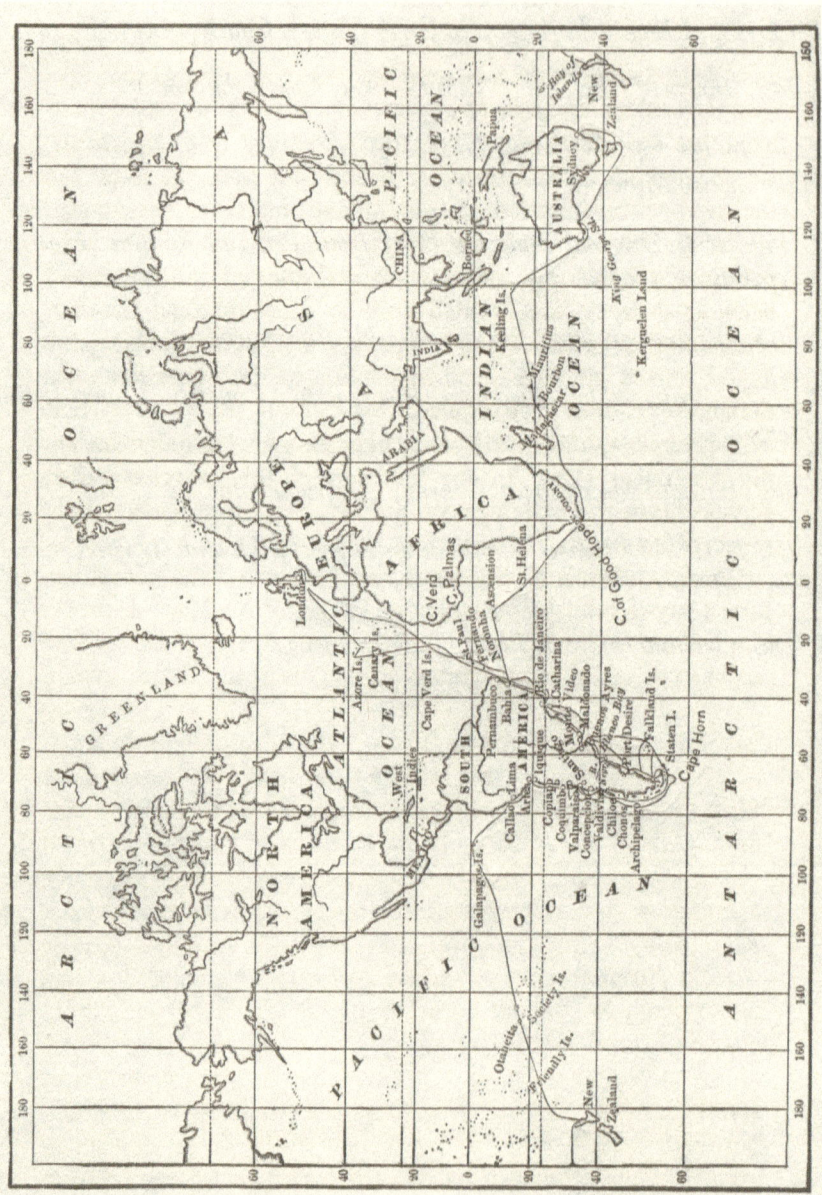

x

ACKNOWLEDGEMENTS

I wish to acknowledge the advice and counsel of many men and women of science, philosophy, and the humanities in the preparation of this manuscript.

I wrote the initial theory in 1991 and 1992, and refined it in the years thereafter. Professors William Ruddick and Roy Sorensen of New York University and Ernst Mayr of Harvard University were helpful during this period with numerous suggestions. My mother, Ida-Louise Ennenga, prodded me with various press clippings, as the specific technologies developed all through the 1990's. She also inspired me with her faith in this work. My daughter, India, kept me off balance by asking, again and over again, flatly, "Dad, what do you do all day?"

Artificial Evolution was published first by MIT PRESS in 1997, as a research theory. Lee Silver, Molecular Biologist at Princeton University and Author, invited me to contribute to the Second Annual Forum on Cloning and Transgenetics, Washington, DC., June, 1998, where I again presented the theory. This conference opened a whole range of new colleagues and friends. Greg Stock, Professor at UCLA and Author, has been an inspiration. He kindly asked me to lend an essay to his collection, "Engineering the Human Germline", OXFORD UNIVERSITY PRESS, 2000. Margreet Jonker, Doctor at TNO Labs, The Netherlands, has been a valuable reader and friend. She has suggested several refinements.

In 1999, I attempted to organize a general reader, which remains unpublished to date. During that process, I became acquainted with numerous scientists who contributed essays and were thoughtful enough to review these new Laws of Evolution. Thanks especially to Freeman Dyson, Prof. Emeritus, Institute of Advanced Study, Princeton, and Author, for his conversations and the thoughts expressed in "Imagined Worlds", published after Artificial Evolution, but confirming its tenets and directives. Thanks also to Dean Hamer, Doctor at the NIH, for his spirited discussions and to Martin Bishop, Director of the UK HGMP Centre, Cambridge, England. All have been helpful.

I want to acknowledge the assistance of the editorial staff at AUTHORHOUSE in the preparation of this book especially Amanda Dollar, Jennifer Slaybaugh and P.J.Rutar.

CHAPTER 1

THE GENE GAME

The New Laws of Evolution

Natural selection, the blind, unconscious, automatic process which Darwin discovered, and which we now know is the explanation for the existence and apparently purposeful form of all life, has no purpose in mind. It has no mind and no mind's eye. It does not plan for the future. It has no vision, no foresight, no sight at all. If it can be said to play the role of the watchmaker in nature, it is the blind watchmaker.

Richard Dawkins,
THE BLIND WATCHMAKER, 1986

I am not a thing, a noun.
I seem to be a verb,
An evolutionary process-
An integral function of the universe.

Buckminster Fuller,
I SEEM TO BE A VERB, 1970

The Gene Game

Prolog

Picture a scene from the 24th Century. Try to slip past the images that first come to mind. Envision a particular place: Something in New Mexico with ultra-modern architecture, or something in Virginia that is submerged below ground, or something 350 miles above Earth, orbiting near our current Space Station.

Now, do you see the same kind of people there that you see today? Do you see the same shapes and sizes? Are they the same creatures that roamed the grasslands of Africa and valleys of Europe 40,000 thousand years ago? Given the advances in technology over the last few decades, is it really possible that our world will not be radically different 300 years from now? Indeed, is it really conceivable that those Humans will not be biologically transformed?

Somehow, instinctively and with sinking reluctance, most of us realize that revolutionary change is inevitable, and already underway. Indeed, technological change is central to our way of life and its speed is accelerating. Some of us fear it, some resist it, some under-estimate it, some get out of the way, while others engineer it, manage it, and expect and enjoy it.

Change and mobility are crucial factors in our personal lives and in the world at large. Change is a universal principle, the heartbeat of nature. Change governs the processes of our planet. All the past geo-physical transformations demonstrate that similar changes will occur in the future. And, just as the physical world was and will be transformed, so too will the life forms, including us.

Recall your basic biology: single-celled organisms, then more complicated ones led to sea creatures that led to land animals that then became more and more complex as each species found its place in life, its own adaptive zone. Remember the dinosaurs that have been portrayed in many films recently, especially the creation-extinction-recreation process of the Jurassic Park Series. The growth, demise, and genomic re-making of species have become common knowledge. Cloning of all types is in the news practically every day. Basic biology is with us everywhere. But, it is getting uncommon, unusual, and even extraordinary.

New developments in biology have captured the attention of the print media, the film industry, made themselves felt in our courts, and in our public and private lives. From conceiving to birthing to dying, biology is headline news, front and center. Every fresh advance creates an issue for more debate. This is so, because the natural processes are no longer taken for granted, to be accepted as events beyond our control. Several generations ago, birth was a simple and risky event that happened in a fashion ages old, its course was accepted, just as was death. Both were mainly beyond human interference or control. Today, we can abort birth safely, we can prevent pregnancy, we can aid it by artificial insemination, or in vitro-fertilization, and we can even examine the genetic make-up of the fertile egg before it is implanted back in the womb. We can view the growing fetus, save and operate on pre-mature babies, and even prevent many inherited diseases. We can engineer trans-genetic sheep, goats, pigs and cattle to produce vital drugs and organs for our benefit. We can create trans-genetic grains to yield higher protein values and vegetables to enhance nutrition and longer ripening. At the other end of the life cycle, we have tremendous powers to modify and extend longevity. All of these techniques challenge age-old beliefs, because they are original to our time. They have yielded many benefits, more fears and confusion, and lots of doubt. Life and death just are not "facts of life" anymore, driven only by eons of biological imperatives. They are now malleable; they are variables.

When we recall our first biology field-trips, and remember those oak-framed exhibit cases showing the slow development of various species, just how one dinosaur or another grew bigger or changed over time, just how the horses galloped across the centuries, we assume a whole set of factors, as stolid as the institutions we visited. Actually, we were taught the verity of those truths of biology. When we think of Charles Darwin and his youthful voyage on The Beagle to those exotic, remote Galapagos Islands, we remember how the animals there influenced his thinking and led to his Theory of Evolution, finally published as THE ORIGIN OF SPECIES in 1859. We know his theory shook the foundations of science and religion and, nearly a century and a half later, it is no less controversial.

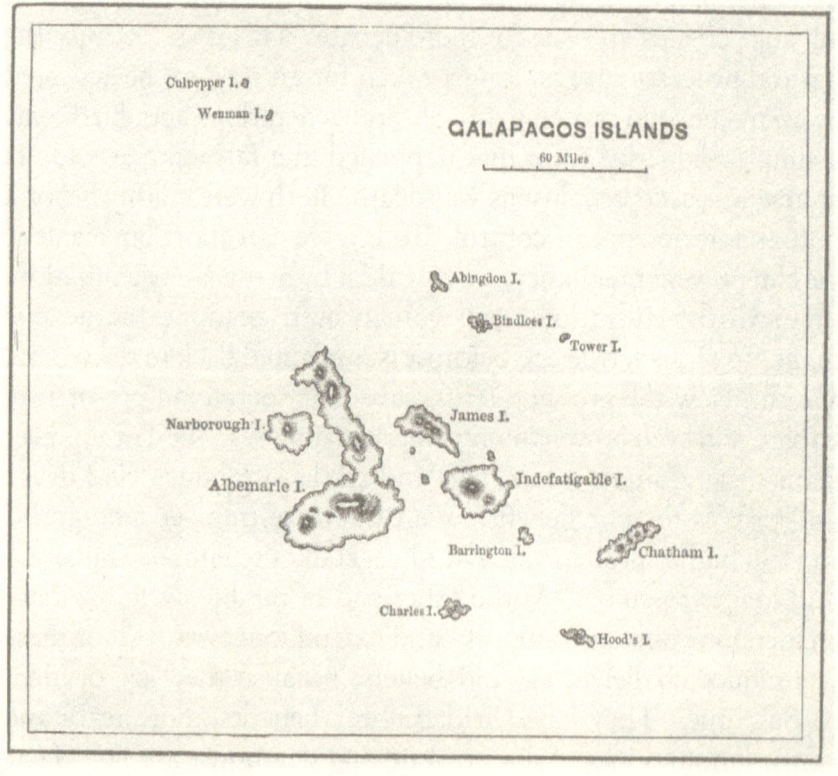

Yet, today, the scientific community, biologists especially, and the public in general consider this basic theory as a given. We assume it is the most natural of natural processes, because it is widely accepted as the best explanation for the development of the natural world. It is the mainspring of nature, the very force driving the natural. Its laws are, like other laws in science, true and immutable, as stout as those oak-frames and solid as those institutions supporting them. And, of course, we consider evolution to be antithetical to anything man-made or synthetic.

Well, like the other biological facts, evolution itself is no longer a given, a fact, or an inevitable process. We have accepted that its laws are unchangeable. We have trusted its constancy. We have viewed the process of evolution like the other laws of physics and chemistry, something as reliable as the Law of Gravity or as fixed as the Speed of Light.

Readers, prepare to shift that paradigm.

* * *

THE NEW LAWS OF
ARTIFICIAL EVOLUTION

Introduction

Darwin's Laws of Evolution have been codified since the 1960's when the discoveries of chromosomal structures were completed. Of course, many workings of the natural world remain mysterious, but the overriding theory of evolution is proven. Most eight graders understand it, and its place, alongside creationism. We have all come to give this theory of evolution status as the force behind nature. We consider it immutable. The law, we think, is as solid as gravity, as absolute as the speed of light. It does give a credible explanation for the origin and growth of species across the world in a simple and comprehensive fashion. It's presentation and consolidation caused a paradigm shift and today it is the widely accepted "world-view". It's principles and its consequences affect our species as it does all others. It is the bed-rock of nature, the source of nature and of our own nature. We all accept evolution as an axiom, as a fixed entity. While it encompasses all the myriad changes in biological nature, the law remains invariable. Truly, Darwin called selection, "natural selection", to distinguish it from artificial selection, or animal and plant husbandry, as practiced by us on domesticated species. This points to evolution and selection as paramount forces of nature. These are the most natural of all natural things, because they are the Laws of nature.

We all separate natural things from man-made ones. We differentiate the realm of nature and the realm of culture or artifact. Even for those who appreciate the contributions we humans have brought to the biosphere, the natural and human are a basic division. Of course, for those who focus on the dangerous and destructive aspects of our species, as the one creating environmental havoc and causing the sixth major extinction, the natural/artificial distinction is razor sharp. It is the scalpel used in many debates, and tool of back-to-nature movements, both past and present. The activist concerns of the 1970's heightened the awareness of this divide. Ever so very naturally, then, evolution secures the natural, and the attributes and principles of evolution are natural. Evolution is the paramount force in living, animate nature.

However, a wide range of recent experiments in biotechnology reveals a new aspect of evolution. The constancy and fixed status of evolution must be reassessed. In the last 10 to 20 years, a series of related bio-processes demonstrate that evolution is now a variable factor. We humans have devised technologies for mainly commercial reasons, but the underlying process is much at odds with the Darwinian natural process. Like aspects of our physical habitat, such as food, water and our immediate climate, the force determining our own development has fallen under our control. The genetic information that directs any organism can now be directed by us. And, the speed of change from generation to generation can be manipulated. Evolution now appears subject to our will and discretion, rather than the other way around. We have created a new mode of evolution. We have created Artificial Evolution.

Artificial Evolution is the controlled micro-manipulation of genetic information from one generation to the next, where the first (variation) step is engineered by us through the choice of and/or transplantation of genes, and then the second (selection) step of survival and continued reproduction is insured by us in a protected environment.

It is a special development of natural evolution; it is the man-made change in the adaptation and diversity of populations. Biotechnology refers to the method, i.e., all the specific technical operations, the chromosomal cartography, the genetic choices and gene transplants, engineering of germ lines etc., while Artificial Evolution names the synthesized process of variation and reproductive survivability as it extends over generations. In many ways, it is an extension of animal husbandry and selective breeding, which Darwin referred to as artificial selection, as opposed to natural selection. Artificial selection takes advantage of the natural evolutionary process. It combines and refines characteristics within one species, usually in a sub-species, while Artificial Evolution can combine characteristics from different families of organisms that could never interbreed. The techniques of Artificial Evolution allow us to precisely manipulate and control that very process on a micro level and, as we shall see, it allows us to halt or slow or even accelerate the rate of change in the process.

NATURAL EVOLUTION

In order to understand where we are headed, we should go back to a sea voyage taken 170 years ago. Picture an erratic line skittering across the globe, like the doodle of an anxious youth. It is the path of H. M. S. BEAGLE that took to sea in 1831. Charles Darwin, a former medical student and then a naturalist and divinity student, was aboard. At 22, young Darwin was included on the voyage because Captain FitzRoy wanted company and hoped Darwin's observations would refute the various evolutionary theories gaining credence. The observed facts upheld something else, something as circuitous yet complete as the path of the BEAGLE itself.

The line of the BEAGLE dropped out of England and swung around the Azores, then slid south to Bahia, Brazil. It hugged the coast to Rio, to Buenos Aeries, looped out to the Falklands, and hopped back around Cape Horn and Tierra del Fuego. The line skipped up the Pacific Coast past Conception, to Valparaiso, Lima, and out to entangle itself in the Galapagos.

It swooped southwest to the Society Isles and New Zealand, then bustled about Australia. It skimmed by Mauritius and Madagascar to Good Hope, South Africa. It angled off Ascension back to Bahia before twisting about itself in its outward course past Cape Verde Isles and returning to the coast of Africa and ending in England. The whole world gone around.

It is difficult to imagine the 1830's world of wholly different parts. Human populations had not yet exploded. The interiors of many continents remained unknown. The creatures were wild and often unafraid. The shores were mysterious. But, more difficult to sense, the globe was not totalized. Electronic media did not stretch to every hamlet. There were no telecommunications; neither cool nor hot media circled the sphere. There were no phone calls; there were not even telegrams. There were no extended rail lines. Even though the world could be conceived as whole to educated Europeans, its parts were disconnected outposts to one another. A letter took months. A mountain range was a major barrier.

Young Darwin went delighted into the jungles of Brazil, across the barriers of the Cordillera, onto the plateau of the Andes, into Australia and South Africa. He examined the mollusks, the ferns, the ancient fossil teeth. He tasted lizards and the soup of young turtles. He greeted the aboriginal peoples of Tierra Del Fuego. Where the path of the BEAGLE looped about the odd isles of the Galapagos, Darwin noted and even rode upon the giant tortoises. He observed the varying beaks of those Galapagos finches that even today, 170 years later, are still being measured, and assessed. He observed, observed, noted and observed. His motto for later naturalists was "leave nothing to memory". He was perhaps the world's greatest naturalist. Yet, even if his observations had come to nothing, his voyage would have been remarkable for its daring and vigor. Although Darwin did not support the creationist views of his captain on the BEAGLE, he deliberated upon his journey, his findings, and his essays for twenty some years before publishing "The Origin of Species" in 1859.

It was a best seller. The educated classes born of the Enlightenment were eager to read and to debate the scientific discoveries of the time. Evolution was a hot topic. The geo-physical and geological discoveries were much at odds with Christian doctrine. The remains of ancient animals really had to be much, much older than 6,000 years, the time line given by Christian calculations. Moreover, the fossil finds were related to current species but not quite identical. The ancient mammoth was not quite an elephant. Some fossils seemed to have no descendants and others continued unchanged. It was perplexing.

There were numerous explanations. While the history of our contemporary theory of evolution is a study unto itself [1], at the time of publication Darwin said that until his time most naturalists believed species to be immutable, created as stable and placed in their settings, but some had begun to see species as changing.

He made a remarkable note in 1836 about the Galapagos: "When I see these islands in sight of each other and possessed of but a scanty stock of animals, I must suspect they are varieties…the zoology of the archipelagoes will be well worth examining: for such facts would undermine the stability of species". Darwin saw that species mutate.

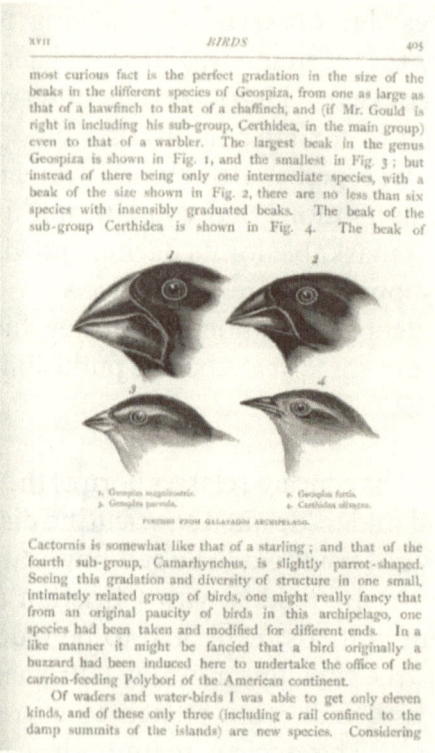

most curious fact is the perfect gradation in the size of the beaks in the different species of Geospiza, from one as large as that of a hawfinch to that of a chaffinch, and (if Mr. Gould is right in including his sub-group, Certhidea, in the main group) even to that of a warbler. The largest beak in the genus Geospiza is shown in Fig. 1, and the smallest in Fig. 3 ; but instead of there being only one intermediate species, with a beak of the size shown in Fig. 2, there are no less than six species with insensibly graduated beaks. The beak of the sub-group Certhidea is shown in Fig. 4. The beak of

1. Geospiza magnirostris.   2. Geospiza fortis.
3. Geospiza parvula.   4. Certhidea olivacea.

FINCHES FROM GALAPAGOS ARCHIPELAGO.

Cactornis is somewhat like that of a starling ; and that of the fourth sub-group, Camarhynchus, is slightly parrot-shaped. Seeing this gradation and diversity of structure in one small, intimately related group of birds, one might really fancy that from an original paucity of birds in this archipelago, one species had been taken and modified for different ends. In a like manner it might be fancied that a bird originally a buzzard had been induced here to undertake the office of the carrion-feeding Polybori of the American continent.

Of waders and water-birds I was able to get only eleven kinds, and of these only three (including a rail confined to the damp summits of the islands) are new species. Considering

That vision is even now more than a little troubling and enormously complicated: Just where does an Eco-system come from? How does it evolve? How do the constituent species interact to change? What factors cause it all? The mystery of mysteries is as poignant now as it was to that naturalist and philosopher, because we are at the heart of that mystery. It is us.

As we shall see, the plot of the mystery is getting a lot more complicated. But, Charles Darwin had his part pretty much right; mainly, his theory was enhanced and elaborated by other scientists over 100 years to become our modern theory of evolution. Although he labored from an enormous set of specifics to arrive at the general theory that rearranged the world, we can accept the theory without reviewing his whole process. The paradigm shift that Darwin produced leads to another, based on the same principles, in our own time, that is perhaps more profound. Yet, all of us who have a passing understanding of evolution can comprehend the process happening to us. A degree in evolutionary biology or biochemistry is not required. An eight grader can get it.

Most of us do know about evolution. We do understand The ORIGIN OF SPECIES. Darwin's work separated theology and creationist thinking from the sciences. After 1860, creationist speculations did not answer scientific inquiries into the age of the Earth, its matter and formation, or its life forms. And, even more fundamentally, Darwin's thought dislodged essentialist concepts about the wider universe. The idea of evolution transformed how we think about our world. It is not a stable fixed entity modeled upon or reflecting an abstract platonic essence or idea; it is underway in change. Darwin let us see the development of life as a metamorphosis rather than a constant.

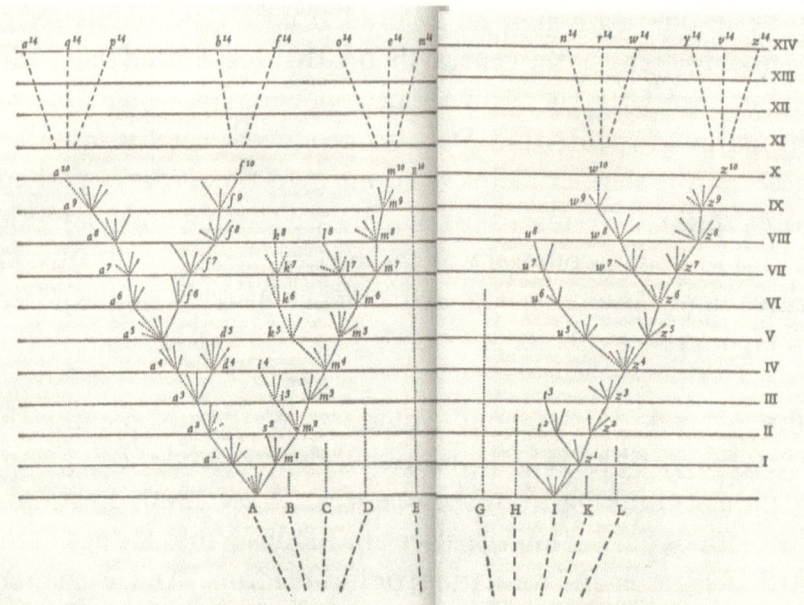

Most of us would say that evolution explains how life evolved on Earth from simple to more complex forms, which multiplied over time and were selected to survive in the changing environments. Actually, Darwin's thought is a bundle of connected theories:

1. **Evolution itself**: that species did change and develop over millennia, along with their zones.

2. **Common Descent or Branching**: that species descended one to another in a branching tree pattern and those lines of development can be followed backward in time.

3. **Gradualism**: that modifications take place over long periods of time as the changes work across the whole species.

4. **Natural Selection**: the force driving the survival of individuals, to maximize their utility and survivability in their changing life zones.

5. **Multiplicity**: that species grow in number to fill the specific adaptive zones of an environment.

This group of linked concepts constitutes the law that revolutionized our thinking about the world [3]. After Darwin, the world and its beings were seen as a process.

Two other major ideas have been added after Darwin to establish our modern theory of evolution. First, Darwin did not fully comprehend the role of sexual selection and genetic re-combination in evolutionary survival, nor did he discover the mechanism of genes. Back then, a major component of contemporary biology, the science of genetics, was focused on mathematical population dynamics. The process of selection was believed then to occur to the unit of an individual, or at the gene site. Thus, the early Mendelian geneticists overlooked the functional sexual individual. Only after the discoveries of Crick and Watson and in the 1960's, was the biological totality of an individual, the genes, the behaviors, the sex appeal and prowess as well as adaptability, recognized as the point of selection. The reality and importance of the individual must be emphasized in biological thinking. [4]

The second refinement to Darwin's theory has to do with the Law of Gradualism. Both Ernst Mayr in his 1954 "Theory of Peripatric Speciation" and Stephen J. Gould in his 1972 "Theory of Punctuated Equilibrium" identify a faster mechanism of species formation than Darwin does. In small "founder" or isolated populations, changes to the gene pool can be carried out more quickly, because there are fewer individuals to affect, and thus, a whole species changes faster. These refinements explained quicker changes in the archeological record and matched observable rapid species modification in founder groups. This notion of speed is critical to comprehending the vast change going on in Artificial Evolution and culture today.

In summary, then, the modern definition of evolution should read "change in the adaptation and in the diversity of populations of organisms".[3] This definition identifies the variation in individuals as well as the whole process of selection on the species level.

It points to the vertical descent in time as new group of individuals is selected in each species and the horizontal effect in every generation of new individuals. If we keep in mind the importance of sexual selection in sexually reproducing species and that the individual organism is always the focal point of selection, then we can be sure in our understanding of this natural force.

So, we have Darwin's Laws. There are several other observations we should make:

Darwin does not propose a divine source for the evolution of life. Furthermore, no scientist or theologian has yet demonstrated this. The Majesty and Divinity of the universe is quite another speculation. That we will take up later. But, Darwinian evolution is strictly biological.

Speciation is not end-directed. As it comes from no outside creator or mind or mover, it has no final goal or destination. There is no process of progressive steps known in advance by an omniscient outside maker.

The process of evolution is disruptive. It is discontinuous. Some species survive; others do not. There have been periods of mass extinction. Some few species continued. Other new species filled the zones of new environments.

Finally and critically, evolution is not progressive. And, it has no value in itself. While speciation gets more complex and manifold, the more complex animals may be thought of as "higher" but they are not better. This is very hard to accept. We flatter ourselves as special. As individuals, we believe in progress, because, of course, in our individual lives we do make progress. Moreover, societies do make progress. However, as a species within the vast evolutionary process, we just happen to be dominant at present. Terrestrial forms are not better than aquatic ones, nor are Tertiary species higher or better than Jurassic ones. And, more species are not better than fewer. From the point of view of the biosphere, the ages on Earth are moments in process, not a progression.

These are the unpalatable and unflattering aspects of Darwinian thought that continue to create resistance and criticism. Not only is it non-theological, it is inhuman and uncomfortable. Nevertheless, these are laws of nature. Darwin is not Homo sapiens-centric. His laws are bio-centric. There is no progress in evolution; there is no necessary continuity; there is no value, other than that held by the species surviving.

With these clear notions in mind, we can assess the adventure we are all on today. Our journey is as harrowing, delightful, and exciting as the BEAGLE. The landfalls are experimental successes, the mountain ranges are ethical debates, the rivers are communal agreement, and the sea storms are social hazards. Our destination, our homeport, is a world transformed, where Darwinian Laws are inverted and those principles reformed, where there is continuity, progress, and value in evolution and where humanity is central.

***

Tierra del Fuego.

The New Laws of Artificial Evolution

1. There is evolution. It is natural, as well as artificial.

2. Artificial Evolution is fast. Changes can, and do, occur in one generation.

3. Descent in evolutionary time is criss-crossed over classes of plants and animals that could not combine with one another in natural evolution.

4. The process of selection is synthetic. We humans select for survival and procreation, rather than nature.

5. There is a growing multiplicity of species, like natural evolution. There are more of them and they are uncommon.

The New Laws of Evolution

The next chapter will show how this new mode contravenes four of Darwin's Laws, leaving only the Law of Multiplicity intact, but enhanced. The human operations of sex selection, genetic scanning, and pre-implantation diagnosis elected to prevent inherited disorders, the technologies of cloning, and the critical, and most illustrative, technologies in animal and plant trans-genetics are all instances of Artificial Evolution. All of these techniques can take effect in one generation. This is very unlike the slow accumulation of changes over many generations. It is much faster than changes in isolated species groups. The Darwinian Law of Gradualism is overturned by these new methods; the old law could not explain these instantaneous changes. Secondly, the Law of Common Descent cannot account for the trans-genetic species that combine genetic information across phyla. The branches are crossed in an unnatural fashion, one impossible in nature. Thirdly, the selection of those individuals surviving is synthetic; we control it, not nature. We decide which trans-genetic plants and animals to clone, breed, or re-seed. So, Synthetic Selection replaces Natural Selection. Fourthly, there is simply more Multiplicity, so Darwin's Multiplicity is Multiplicity Plus. And, finally, the Law of Evolution itself is modified, because while there continues to be modification of species, its mode is wholly and functionally distinct. We now live in Artificial Evolution.

Darwin's Law	The New Laws
1. Evolution	1. Artificial Evolution
2. Gradualism	2. Immediacy
3. Common Descent	3. Trans-class Descent
4. Natural Selection	4. Synthetic Selection
5. Multiplicity	5. Multiplicity Plus

Natural Evolution	Artificial Evolution
Natural Selection or artificial selection	Synthetic Selection (gene manipulation and survival)
Common Descent	Trans-Class Descent
Gradualism	Immediacy
Multiplicity	Multiplicity Plus

Artificial Evolution then is not merely a change in degree, but rather is a change in kind. It is different in substance. First, Artificial Evolution is based upon the ability to rearrange the genome of an organism by changing genes or exchanging them in trans-genetic operations with other species in an exact way. This precision was never achievable before in animal husbandry; the techniques have only been devised in the last 20 years. Secondly, organisms are created by combining genetic material across classes. Thirdly, Artificial Evolution can control the rate of genetic change from one generation to the next, which selective breeding cannot do. So, it is the precision and trans-species exchange, as well as the control of rate change which is achieved immediately, that are the special characteristics of Artificial Evolution.

Natural Evolution, of course, is a two-step process, of genetic variation and then selection of individuals best fitted to a changing environment. In the first step, the unique combination of genes is formed in the individual, but the individual is chosen and selected in the second step by surviving and reproducing in its particular environment. In Artificial Evolution, the first variation step is engineered or controlled by us, and the second selection step is not so much controlled as guaranteed by us. The individuals are selected by us, and their further reproduction devised by us. This is synthetic selection. Without the oversight of humans, the second step would be as dependent on natural conditions as in natural evolution, but then, so too would be the first step. In other words, in Artificial Evolution, humankind executes both steps of the evolutionary process.

The specific operations, that is, the biotechnology of Artificial Evolution rely on the same chemical and physical phenomena as natural evolution and, indeed, confirm them. While Artificial Evolution is qualitatively different from natural evolution and while it contradicts the Law of Gradualism, Common Descent, Natural Selection, and Evolution, this does not mean that it overturns Darwinian Laws in natural evolution. To the contrary, modern Darwinian Theory continues to explain the natural environment.

For the first time in evolutionary history, we can manipulate evolution and do it immediately. We can now control the random mutations, which give rise to variations in individuals of a population, and thus provide for the survival of a species under changing environmental factors. An immediate reorganization replaces the gradual process of change. It is true that mutations would continue to occur in individuals in this new mode, and at the same rates as naturally. While most would be "neutral", that is, they would not be noticed or raise concerns in most observers, any and all of these mutations could be scanned and manipulated into the following generation. Of course, as a practical and technical matter, controlling the genome of every organism on the planet would be absurd. Legions of bio-technicians diagnosing endless pre-embryos would still quickly fall behind the task. As stated above, Artificial Evolution is an offshoot or special case of evolution; it is only in particular cases that we can or want to manipulate change in a species. Still, for the first time, the gene flow from generation to generation is under our direction; speed of change and flow from generation to generation can be arrested or accelerated. The point here is dramatic: Humanity is no longer entirely subject to the process of natural evolution.

If various genetic operations were only isolated cases, or if each operation were restricted to an individual, the specific instances could not substantiate a generalization called Artificial Evolution. In other words, restructuring the phenotype rather than the genotype of an organism, or altering one individual rather than the species, could not be categorized as a new mode of evolution. However, cloning in agriculture is widespread, effects large parts of some species, and could be used more widely on large mammalians, as research shows. Although in-vitro fertilization and pre-implantation diagnosis are now limited to preventative operations for a few, it is another indication of our control of evolutionary processes. And, transgenic experiments and patents are more numerous. They create many individuals that could be called sub-species, and, clearly, they affect the genotype of the organisms involved. Thus, although this new mode of evolution is an offshoot of natural evolution, the instances and practices underway today do constitute a general process, one that is substantially distinct.

To summarize the eight characteristics of Artificial Evolution:

1. Evolution can be controlled, manipulated, structured, and synthesized by man. Artificial Evolution selects desired genetic material and manipulates it from one generation to the next. In Artificial Evolution, the genetic time clock is wound and set by humankind.

2. It occurs in an immediate, not gradual, fashion. It contravenes Darwin's Law of Gradualism, Mayr's Peripatric Speciation, and Gould's Punctuated Equilibrium.

3. Subsequent generations can be similarly manipulated, scanned, corrected, copied. While mutations will continue to occur naturally, even though most would be neutral, the process can correct for then.

4. Trans-genetic plants and animals are created by combining genes across classes in a process that nature could not have managed and that is distinct from Darwin's Law of Common Descent.

5. Without our active oversight, natural breeding, birth, and natural evolution would recommence. But, we do select and

insure survival. This synthetic selection is very different from natural or artificial selection.

6. In cloning, identity is the main characteristic, not diversity. Cloned forms lack the characteristic that Darwin identified in his evolutionary theory, "every individual a spontaneous generation." [2] Just as Artificial Evolution is a unique event in natural history, so too are its life forms unprecedented. Clones, as well as trans-genetic beings, are a new form of life on earth.

7. For the first time in history, humanity has created a new kind of living thing.

8. Finally and significantly, Natural Selection is no longer the only directional force in evolution. There is a new mode of synthetic selection. There is a new mode of evolution.

We can move now to speculations on Artificial Evolution as an event in natural history. Molecular life has been in natural evolution for 3.5 billion of our earth's 4.5 billion years. A mere one hundred and fifty years ago, Darwin initiated his thought on the dynamics of evolution and common descent. Only in the last century has biology become a separate science whose principles have been identified as specific and distinct from the other sciences. Only in the last forty years, have the basic mechanisms of inheritance come to be understood. Now, in the last twenty years biotechnology has given us Artificial Evolution. It would seem that we have created Artificial Evolution to control that very process that brought us forth and determines our function within our habitat. Just as we have sought to overcome the forces of nature, and determine the factors of our habitat, we have now managed to achieve dominion over the source of our arrival and development—evolution itself. Evolutionary history is diverging, as though our artificial mode were a loop out of evolutionary time. This loop can extend, grow arabesque, and complement or foil the natural evolutionary time line. And, the quality of the loop, the thickness and turn of the line will come from us, from the hand of humanity.

Since humans are attempting to control the very process of our coming, it would appear that we are trying to evolve out of evolution. Because evolution has determined us and given us the capabilities to control the process, we are able to get beyond it. Artificial Evolution is our way of evolving out of evolution.

This book argues that our species has profoundly disturbed the planetary relationship of Sun Source/ Plant/ Animal. The workings of the earth are badly out of balance and we must create a new harmony. The situation is so far beyond repair or conservancy, that, if we wish to avert monstrous global catastrophes and mass extinctions including ourselves, Homo sapiens, we must pursue another direction and we must do so in this century. It is senseless to judge any species for their natural behavior. We should not blame ourselves for employing our intelligence, organization, and cultural creativity in dominating the biosphere. Rather, we should utilize these genetically evolved capacities to reorganize and create a new type of equilibrium in a humanized sphere. The thesis is a new vision of a sapiens-centered solar system.

Most doubt the wisdom and ethics of this undertaking, others fear it, and some just reject it automatically for religious or cultural reasons. The argument of the chapters ahead will show that we are on an adventure more momentous and challenging than young Darwin's on the BEAGLE. The turn of our new evolutionary line will twist across time as the path of the BEAGLE circled the globe, and as Darwin did, to arrive at another new vision of the world, of its evolution and its human purpose. Our new line will draw us into evolutionary continuity and value, and ever closer to the dynamic natural principles of change and mobility.

References:

1. Mayr, Ernst. Titles 2 & 3 cover this synthesis in Evolutionary Biology.

2. Mayr, Ernst, Growth of Biological Though, Harvard University Press, 1982, Darwin Notebooks, 132, quoted p. 493.

3. Mayr, Ernst. Towards a New Philosophy of Biology, Harvard University Press, 1988, p.163 and p. 186.

4. Ibid, pp. 196-210.

NOTE: Artificial Evolution

This term should not be confused with others in the literature, with evolutionary synthesis theory (Huxley, Mayr, et al 1935-1947), with social ones (Spencer, 1884), or with progressivist ones (Huxley 1942, de Chardin 1959, Anderson 1987, Wesson, 1991). Nor should it be confused with germ line engineering or reprogenetics (Silver, 1997), which are examples of Artificial Evolution.

CHAPTER 2

Bio-Technics Of Artificial Evolution

Bah, Ram, Ewe!
Bah, Ram, Ewe!
To your breed,
To your fleece,
To your clan,
Sheep be true!
Bah, Ram, Ewe.

The password that the sheep gave to Babe, so he, the little pig, could win the sheepherding contest with the Boss.

BABE, MCA Universal Pictures, 1995.

Bah! was the word heard around the world when "Dolly", the clone made headline news in 1997. Ian Wilmut of the Roslin Institute in Edinburough Scotland announced this first creation of a cloned mammal from another adult to great fanfare and shock. The telephone rang constantly day and night for a week. Even the phone in the janitorial closet was ringing for news briefings. Was "Dolly" true to her fleece, her clan? What is it to be a sheep and, to be true? Indeed, what is it to be a clone?

The Roslin Institute did not set out to create headlines or controversy. Actually, the Institute did not even want to make a clone for its own sake. "Dolly" was the solution to a scientific problem in bio-engineering proteins for therapeutic uses. There are 50 to 75 of these proteins in bio-technological production. Some are for Hemophilia, some for blood clotting, like "ATP" and "C" proteins, some for Cystic Fibrosis. All are in development because they can be made less expensively in animal processes than any other way.

The process is transgenetic. The chromosomal database, or gene sequence, that will produce a particular enzyme in one species can be clipped into the gene sequence of another species, transferred from one species to another, even across classes of organisms. This unlikely process was first done in bacteria in the 1980's to produce special enzymes, and then in mice to study immune systems, and then worked up into larger mammalian forms, so that sheep and goats were bred with these transgenetic additions to secrete human blood-clotting proteins into their milk. When the milk is evaporated, the therapeutic protein is left. It is then marketed for human use at a much lower cost. But, in making these transgenetic sheep, the eggs must be manipulated, the nuculi removed, the new gene sequence added, the data replaced into the egg and then implanted into the ewe to grow to term. There are many steps to go wrong and, of course, they do. Many eggs are wasted, and many implantations are unfruitful. So, Dr. Wilmut and the Institute conceived of cloning from a grown proven adult.

But, until they attempted it, cloning from adult to adult was deemed impossible. And, this was the scientific breakthrough and marvel of "Dolly". They succeeded.

"Dolly" was produced by taking cells from the udder lining of another adult sheep, extracting the genetic material from the cell, and then introducing the same material into a fresh neutral egg, shocking it with a zolt of electricity and implanting it into a ewe to grow. There were miscarriages. There were defects. But, finally, there was "Dolly", curiously oversized and with tissues the same age as the original adult. The process works from other cells as well, from breast and skin cells. Thus, an exemplary adult could give hundreds of cells and nuculi for cloning. In theory, then a particular adult could be cloned many, many times into a herd with the identical genetic characteristics. To comprehend the significance of this, you must realize that the drug market for Hemophilia is $1.5 billion annually, and this can be supplied from a small herd of Heffer cows in milk production. About 40 cows can supply the total world market in a tidy, scientific farm. This is typical of the economic incentive for mammalian cloning.

DIAGRAM OF THE NUCLEAR TRANSFER PROCEDURE THAT PRODUCED

Eggs recovered

DNA removed from egg

Udder cells placed in culture

Transfer of cell

Egg and cell fused with electric current

Egg cultured (7 days)

Cells divide to form a blastocyst

Egg transferred to Blackface surrogate mother

Birth given to lamb (Dolly)

Three genetically identical Brangus bulls that were produced by the Granada Corporation of Houston.

So, "Dolly" became infamous for all the wrong reasons. "The First Clone", read the headline news. She was the first clone from an adult and very important for that. The whole process is highly significant for commercial and theuropeutic reasons. But, "Dolly" was far from the first clone. As early as 1954, frogs were cloned [5]. The eggs are easy to manipulate and the process simple to understand. As the initial fertilized egg divides into a mass of 32, then 64, all the cells are undifferentiated and identical. Each cell can be separated and grown to term, so the resulting tadpoles and adults are the same genetically. They are identical. And, they are *almost* like 2 identical twins. The process here just separates the growing cell mass before differentiation to create the clones.

The agricultural industry adopted the process in the 1980's in an effort to idealize and standardize its products. In 1988, Houston-based Granada Corporation produced seven Brangus bull clones, all identical siblings [6]. Here, as previously done on other organisms, the genetic material from a chosen 32 cell "pre-embryo" is extracted from each and inserted into fresh eggs whose nuclei have been removed. The new identical eggs are then implanted into surrogate mothers, and the siblings are born as identical clones, not twins. And, indeed, this group of identical bulls has real authority. (See Figure A)

This was the first large, mammalian clone and gained much attention for Granada Corporation and its cloning effort for herd characteristics. With the process of cloning from adults, rather than from eggs, obviously large herds could be developed with desirable champion profiles. This is one of the most remarkable facts of bioengineering. It also illustrates a unique aspect of this new mode of evolution.

The individuals in the set of clones are genetically identical and comparable to a single celled organism scattered across the habitat. They may be likened to multitudinous twins; they are units. The Brangus bulls are, moreover, exemplars of their form. It is as though an Aristotelian form of nature, "The Bull", came to life. Identity, not uniqueness, is their main characteristic.

Polly and sisters, each carrying an additional gene that produces a clotting factor in their milk

Clones and cloning are topical and curious. "Dolly", and her clan, certainly got attention; in 2004, a group of Mules were cloned from an adult in the same manner. But, they are all instances of a broader process. A clone is the opposite of the natural process of uniqueness. A clone is *not* like Darwin's insight, "every individual a spontaneous generation." The whole process of cloning is done to maintain characteristics in copies of the original. It is especially poignant in large mammalian animals, and, although they can be compared to identical aspects of twins and less complex life forms, they are unprecedented. They come from the new mode Artificial Evolution, and illustrate its main characteristic: We control the genetic flow.

Cloned-beings, then, are a new form of life. They are bio-technically produced, identical in large numbers and can be identical for generations. It is as though the genetic clock stopped. Change is suspended. Cloned beings live in an arrested gene-flow. Although site mutations continue, generational recombination and mutation is halted. The central reason for the sexual recombination of gametes in nature is to provide variability to respond to changes in the environment and clones are distinctly the opposite. We have created a new kind of being. Identity, not uniqueness, is the main feature, and the gene flow is in stasis.

Trans-genetic animals epitomize Artificial Evolution. By introducing the genetic information of one species into another, new life forms are being created and patented. The genes that cause specific enzymes to be produced are being added to sheep, cattle, and pigs, as well as insects, fish, birds, and amphibians. [4] Pharmaceutical Proteins Ltd., a U.K. firm, has produced trans-genetic sheep that secrete in their milk Alpha—1—antitrypsin, an enzyme used to prevent mucous from building up in the lungs of people with Emphysema. Genzyme Corp. of Cambridge has trans-genic goats that give in milk TPA, tissue plasminogen activator, which dissolves blood clots and is used to clear blocked arteries in heart attack patients. Others are a transgenetic bull from Genpharm International that gives human lactoferrin and transgenetic pigs from DNX Corp. [1] of Princeton, N.L. that give human hemoglobin. These follow upon the many transgenetic rodents developed for testing and the trangenetic plants produced over the last decade, such as cotton to resist caterpillar's pests, potatoes to resist beetles, and tomatoes to delay spoilage, among the 50 species affected. [2] Anti-coagulant genes from rabbits are being added to mosquitoes, so the life cycle of malaria, and thus its spread, might be controlled. Every year new transgenetics are created. All are examples of micromanipulation of genetic material from one generation to the next, by transplanting genes from one species to another, so as to create animals unnatural, super-natural, or ultra-natural.

Unlike the gradual step by step accumulation of changes that Darwin identified in natural evolution, all these strains occur immediately, in one generation. The reorganization of the genotype occurs in one biotechnological operation, rather than over generations. This immediacy is one of the most distinctive characteristics of Artificial Evolution.

Darwin's Law of Gradualism holds that species change over long periods. Despite recent modifications in that analysis by Mayr and Gould, the process of transgenetics creates new sub-species immediately in one generation. So, this is the first of Darwin's Laws that is different. The Law of Immediacy replaces Gradualism.

The transgenetic plants and animals reveal another difference with natural evolution. In nature, species have descended along lines of development in fitting into their adaptive zones. Their genotypes are related in a linage that can be traced backwards to where branches or new lines of speciation occurred. Darwin recognized this process as the law of Common Descent or Branching. When a DNA sequence yielding anti-viral enzymes or anti-coagulant proteins is placed into the genes of a different organism, when a sequence of a rabbit goes into a mosquito, when the gene yielding omega 3 fatty acids from salmon is added to pigs to make the meat healthier for human consumption, or when human hemoglobin producing sequences are clipped into pigs, the descent is not common. The branches here are crisscrossed. None of the transgenetics could have occurred in nature or ever would have; nor could they be traced back according to the Law of Common Descent. So, the Law of Transgenetic Descent replaces the second Darwinian Law.

In the chapters ahead, the various issues of transgenetic species will be examined. Biologically, they are fascinating. Of course, they are oddities and compelling as such, but they raise many questions about their relations to their founder species: how they can be classified? how their traits might be breed back into natural species? how their ownership is held? And, how they can be patented and regulated? In short, there are complex ethical, economic and social issues, as well as scientific and taxonomic problems. But, there are numerous therapeutic and medical benefits, and, quite possibly, environmental benefits to endangered species. Certainly, however, the transgenetic plants and animals live and develop under Laws distinct from Natural Evolution.

Just as with clones like the bulls and sheep, transgenetics survive into the next generation under the supervision of humans. We copy again an adult to another clone, or manipulate the genes to produce another transgenetic, or mate the transgenetics to insure that the traits are carried forth. Clearly, this is not Natural Selection, because the individuals are not succeeding or failing and sexually reproducing in a natural zone. Nor, is it Artificial Selection, because we humans are not observing and selecting traits and behaviors to be combined in sexual reproduction. Transgenetics are created by gene splicing, by pre-embryo analysis, by in-vitro fertilization, by synthetic implantation, and their survival is protected, their procreation guided by us. Another Darwinian Law is replaced; Natural Selection/ Artificial Selection becomes Synthetic Selection.

We must remember that the techniques are examples of this broader new Law. They have created the Law and are its instances. There will be more techniques, and the new Law will become clearer. It is especially important to recall this difference in the human technologies, because we tend to focus on the immediate operations, rather than the general laws at work.

The biotechnologies affecting human beings confirm these new Laws and demonstrate them in other more subtle ways. And, of course, human engineering excites more ethical and social questions. All of the procedures in practice today are elective surgeries undertaken for theuropeutic reasons, mainly to avoid or correct hereditary disease. All are closely monitored under ethical codes of medical practice. [7]

A great impetus for bioengineering and for the Human Genome Projects came from the detection of the gene sites for Huntington's disease during the 1980's. First one family, then another and another, were analyzed to demonstrate that the disease was inherited. Considerable luck led genetic scientists to detect the locations on the chromosome that yielded the proteins causing the malfunction in victims. Huntington's disease led the way. The same method continues: family patterns are examined to determine if a disease is hereditary and then the search for the critical gene sites moves forward. This search indicated the need for a full map of the Human Genome that was underway at the NIH and Celera Corp. for a decade and completed in format in 2000. Almost 2,000 inherited disorders have been identified and thousands of gene sites are associated with the diseases. Now, the fertilized egg mass of a pre-embryo can be examined for deleterious genes, and only a disease free egg is used for implantation into the mother, thus eliminating the inherited factor. While this is laudable and right, the natural evolutionary process has been manipulated and restructured by us.

In the human fertilization labs of today this preventative method is practiced on numerous inherited disorders that have gene locations identified. Hemophilia and Duchene's muscular dystrophy are sex determined diseases, and so, pre-implantation diagnosis is used to select for sex, thus insuring that when the pre-embryo is implanted, the fetus will be unaffected. Cystic fibrosis stems from abnormal genes, and so, only a pre-embryo with the normal genes present is chosen for implantation. [7] All are corrective measures whereby the genetic material is controlled into the next human generation. In England, X-linked mental retardation, adrenoleukodystrophy, Lesch-Nyham Syndrome, and well as DMD are being avoided by pre-implantation diagnosis and sex selection [3].

With the complete map of the 32,000 or so active genes on our chromosomes available, the real theuropeutic work can begin in earnest. Huntington's disease, Tay-Sachs, that cripples and kills children, Cystic Fibrosis, and PKU are some of the cruelest and most devastating maladies, and especially, because they are inherited, predictable and, until recently, inevitable. We now have the power to make them part of the past. So, the preventive interventions in the human biotechnologies demonstrate the same Laws of Artificial Evolution. The genetic material is micro manipulated one generation to the next; it occurs by Immediacy; it is done by Synthetic Selection.

All of the animal and plant clones and all of the transgenetic sub-species are unprecedented here on Earth. We humans have created new forms of life. They all come from Artificial Evolution, and even though this mode is evolutionary, it is distinct from Natural Evolution. Thus, the fourth Darwinian Law, Evolution itself, is transformed. In its place is Artificial Evolution.

The recent cloning of monkeys alerted most observers that, indeed, speculation about human cloning had better be taken seriously. Indeed, there has been one false claim already. While it has been

39

discussed for years, practical success in a closely related species proves that human cloning could be done. Why and when are the difficult questions. Yet even now, cloned humans might be undertaken at the same time as transgenetic humans. Today, gene sequences for infrared vision or the genes for sonic echoing from bats might be added to humans [8]. An enormous reformation and redesign of human nature now appears quite possible.

This bioengineering of humans is called engineering the human germ line, or genome [9]. Reprogenetics is the term used by Lee Silver [10]. We have just reviewed passive engineering by sex-selection for inherited gender disorders. The selection of unaffected pre- embryos is also passive engineering. When we substitute gene sequences or transpose genes from other species we would be actively engineering off spring. This is the same transgenetic process as in plants and animals, but applied to human beings. This engineering, or reformation of the genome, or reprogenetics, is an instance or technique of Artificial Evolution and functions by its Laws.

The most likely development of these procedures will commence when typical American families begin examining their potential children for inherited disorders. The next step will be examining and choosing behaviors, and then aptitudes. Mothers will ovulate dozens of eggs through hormonal stimulation. The father's sperm will then fertilize them all. These will be incubated several days in vitro until each pre-embryo has 6-10 cells and a sample nucleus can be examined for its specific genetic make-up. The information will then be displayed so the parents can choose which of their possible children to bring to life. This whole process takes place in a petridish, outside the mother and long before implantation in the wall of the womb. After the selection and scanning for diseases, attributes, and traits the pre-embryo will be implanted in the mother to grow. While this process is severely constrained at present, within fifty years it may be in full practice [11].

The notion of adding genes with unique characteristics is an addition to the process. This would be actively creating new humans by transgenetic operations, whereby the chosen pre-embryo would be modified with an additional gene sequence. Practically this is also fifty years away. Again, all these technologies are instances of our new mode of evolution, just like those practiced in other species.

* * *

Let us restate the definition and Laws of Artificial Evolution:

"Artificial Evolution is the controlled micro-manipulation of genetic information from one generation to the next, where the first (variational) step is engineered by us through the choice and/or transplantation of genes, and then the second (selection) step of survival and continued reproduction is insured by us in a protected environment".

IMMEDIACY	Not Darwin's **GRADUALISM**
TRANSGENETIC DESCENT	Not Darwin's **COMMON DESCENT**
SYNTHETIC SELECTION	Not Darwin's **NATURAL SELECTION**
ARTIFICIAL EVOLUTION	Not Darwin's **EVOLUTION**
MULTIPLICITY PLUS	Not Darwin's **MULTIPLICITY**

This leaves only Darwin's Law of Multiplicity unchanged. This says that species multiply over time to suit and specialize in new adaptive zones. This Law remains in Artificial Evolution. It will be confirmed by our creation of many new life forms, super-strains, ultra-species, and un-endangered species. We humans will multiply the multiplicity. Multiplicity only begins to name the daring adventure out of evolution that our species has undertaken. It is Multiplicity Plus.

We now know what a clone is, and we now understand this new mode of evolution. We began this chapter with the questions: was "Dolly" true to her fleece, her clan? What is it to be a sheep and to be true? Indeed, much more broadly, what dignifies life? What are the ethics of this new mode of Artificial Evolution?

References

1. Elbert K., Denman J., Krimpenport P., Wright G. 1991. <u>Bio/ Technology</u>, September.

2. Gasser C. and Fraley R. June 1992. "Transgenetic Crops", <u>Scientific American</u>,, p.62.

3. Handiside, A.H. April 1990. "Pregnancies from Human Preimplantation Embryos Sexed by Y-specific DNA Amplification". Science Vol. 344, p. 768.

4. Murphy, D. 1999. <u>Transgenetic Animals in Agriculture</u>. Cab Press.

5. Zimmerman, B. 1984. <u>Biofuture,</u> Plenum Press, N.Y.

6. New York Times. February 17, 1988. "Cloning brings Factory Precision to the Farm".

7. Interview. October 1991 "Center for Reproductive Medicine", Cornell University, New York City.

8. Discussions. 1998. "International Conference on Cloning", Washington D.C.

9. Stock, G. and Campbell. 1999. <u>Engineering the Human Germline</u>, Oxford University Press.

10. Silver, L. 1997. <u>Remaking Eden</u>, Avon N.Y.

11. ----------------------, p. 200ff.

CHAPTER 3

Ethics

"Species are groups of interbreeding natural populations that are reproductively isolated from other such groups."

"Every species is the product of evolution, of speciation."

"The isolating mechanisms of a species are a protective device for well-integrated geno-types."

ERNST MAYR [1]

"The human being must be respected—as a person—from the very first instant of his EXISTANCE".

INSTRUCTION FROM THE VATICAN, 1998. [2]

Was "Dolly" true to her clan? What choice did she have? Who decides? And, what if Dolly were another animal, say, a human being?

The ethical debates concerning biotechnologies are extensive [3]. Each individual procedure is questioned, analyzed, and discussed from many points of view. The animal and plant procedures get less attention then those effecting humans, but the accumulated redesign of so many plants, like high protein rice, disease resistant corn and soy-beans, or long shelf-life tomatoes, and animals, like fast growing farm fish or high milk producing Friesian cows, is gaining greater social awareness. Most of the arguments are specific to each practice. Very rarely does an ethicist formulate an overall ethical system upon which each process is judged. Consequently, each technology gets multiple reactions, rendering an overall muddled disagreement.

The human technologies of birthing have received the most extensive debate. Artificial insemination was first ridiculed as "turkey-baster babies"; in vitro fertilization was ridiculed as "test-tube babies"; genetic engineering as "designer babies". Surrogate motherhood was hotly disputed from the beginning, especially because the legal status and emotional relations of birth mothers and biological mothers with the children is so complex. The therapeutic practices of pre-emplantation diagnosis on intro-vitro human zygotes have been discussed at less length. Gender selection to avoid sex-linked inherited disorders receives less resistance, because these therapeutic operations do such good. However, the techniques in active gene modification of the human genome, or notions about transgenetic humans, are totally rejected on some accounts and suspicious on others. Pejorative words like "mutants" are immediately used, when of course, all species, including humans, mutate. Then, the practices of cloning excite wild reactions, despite again the natural process of identical twining, to which cloning is closely related.

The general picture is an array of practices that are reviewed separately from one another and analyzed most typically without ethical systematics. Various individuals give opinions about separate bio-techniques, usually in the pop-press and in sound bites. Often the creative scientists have differing opinions with even less methodology. The overall color of this picture is a mucky gray of orange-purple-green mixtures. Curiously, this confusion may explain the lack of alarm or acceptance from the wider public. A broader presentation of information is required, along with a comprehensive ethics.

We have said that all of these biotechnologies are instances of Artificial Evolution. Each procedure that affects the genetic make-up of the next generation, and its survival, demonstrates this new manner of evolving. Indeed, these practices take place within Artificial Evolution. So, we have piece-meal debates on the process of egg and sperm collection, in vitro fertilization, pre-implantation diagnosis, gene modification, gender selection, transgenetics, and cloning, all as a disarray of opinions with little ethical systematics. We must remember that the techniques are examples of our new evolution, actions in the gene game, and we need an ethics for the whole process.

Artificial Evolution is really the canvas upon which this grisaille of opinions hangs. Perhaps it is also the frame. Since this substrate and structure has not been presented as a whole, of course, there is no history of ethical debate on this mode of evolution itself. The parts have been seen, and colored a less than sensuous gray, but the picture has not been unveiled and its support and framing have not been appreciated. Furthermore, its hanging and home setting have not yet been perceived or devised. Once this shift in the Laws of Evolution has been credited, then it's setting within other physical and chemical Laws may be appraised. The immutable Laws governing our universe may become malleable just as our evolution has, so the placement of this new picture in a broader setting has considerable significance. Our ethics about Artificial Evolution can serve as a guide and base upon which future modifications to the Laws of Science may be assessed as well. So, a systematic ethics is required.

Since Artificial Evolution comes from us and certainly stems from technocratic and scientific endeavors, it clearly reflects an anthropocentric or sapiens-centric, rather than a biocentric viewpoint, for what other simulation of a natural phenomenon could be more poignant, what other human intervention more powerful. Few will not fret over the long-term effects of this new mode of evolution. Just as our species has profoundly affected the natural world through cutting the forests and developing agriculture, through annihilating many species and domesticating others, and through generally reforming the global environment to fit our wishes, so Artificial Evolution will extend our dominion. It will enhance our power to design the living environment.

The transgenetic plants and animals of today may well prefigure super-strains developed for specific environments. Presuming knowledge of the genome of other species, we might move well beyond the transgenetic animals of today to create organisms that bridge phyla and perform in unusual environments. We might greatly increase food production by redoubling efforts on transgenetic plants. Or, we might shift the evolutionary line of some endangered species to make them more compatible with the increasingly humanized global environment, for conjectural example, by boosting the egg yields of Peregrine Falcons or Condors, by changing the feeding habits of endangered cats, or by down-sizing the territorial requirements of numerous species.

Although every particular procedure need not be discussed, there are a number of current hot-topic issues that can be addressed. Was "Dolly" true to her clan, as we asked? We know what a species is, we know that transgenetics do not disrupt the species unity, and that they do hand down their unique genes. But, what gives humans the right to intervene in their genome? Or, any genome?

Historically, we have assumed the right to domesticate, breed, and use animals and plants. Darwin called this Artificial Selection. Our right is based upon our appointed superiority as a species. To the degree our claim is well founded, we have made transgenetics with the same authority. The breeding and spread of these unusual genetic combinations is easily controlled in the domestic mammals. Fish, birds, and especially insects are more prone to escape into wild populations, so their captivity is closely guarded in double fail-safe facilities. Their uses and release into nature is carefully studied [4].

Human cloning is related to identical twining, but, if it were just the same sort of thing, then the reactions would not be so dramatic. Most people, including the lead scientists, oppose it instinctively, because it seems unnatural and unnecessary, and then back up their reaction on the sanctity of life grounds. Yet, there may be some compelling reasons to consider the procedures for human beings. Pre-embryo fetal cells might be saved for later use by individuals to replace organs. Cloning from adults would allow infertile couples and homosexual couples to have a child directly from one, and then another of them. Although there are other means of procreation, these would be special situations. Pre-embryo fetal cells might also be saved to clone a child lost in a tragic accident; this could be significant for single child families. While there certainly is oddness about it, both European and American supervising committees have approved use of pre-embryo undifferentiated cells [5]. Although there are many cases for considering human cloning, it is rather the long-view in combination with gene modifications and additions that pose the critical possibilities and questions.

Of course, this brings up a substantial set of ethical issues. The distinction of Human/Nature and artificial/natural is crucial to the matter. Some contend that the natural environment must include our species and our effects upon the natural. Thus, the natural world has been and will continue to be restructured through our interventions: We are nature too. Others have us outside the natural order, for various reasons and towards even more various ends. This derives from long religious and philosophical traditions that give us Divine causation, a continuous state of speciation in a constant world, Humanity's elevated state in the "Scala Natura", and a cosmic teleology [6]. Still others separate our species from Nature on the empirical evidence of our ability to reason, our social structures and our invention and use of tools. Most of the ethical positions concerning biotechnologies rely on one or another of these premises.

If we take Nature to include Humankind, then our new mode of evolution is another adaptation of natural processes and we could call it *evolution squared* or some such thing. But, this would only serve to rename the distinctive characteristics. There is some ethical absolution if we dispose of Divine oversight and cosmic teleology, because it can be argued that Humankind is merely intervening and forming the environment as it has and will. Another line of argument from the same assumption, a tougher one, would hold that our natural capacity to organize and dominate should be matched by our natural ethical oversight within the biosphere.

However, the admission and recognition of the dichotomy of Humanity/Nature will fire the ethical debate about Artificial Evolution. Although there is no body of debate about the Artificial Evolution as yet, an overview of the existing debates on specific biotechnologies will indicate the formats for Artificial Evolution, because the values involved are charged and fueled by the same long traditions. The ethical debates fall into several groups, and curiously unite people from the political–religious right with those from the political–environmental left. In these ethical debates, it is important to remember the Congressional Act of 1988 that prevents the patenting and sale of human genetic information. There are oversight committees in agricultural development and animal husbandry and, generally, the ethics of scientific research and medicine are practiced keenly today.

The first group of objections comes on religious grounds. The Vatican's "Instruction on the Respect for Human Life in its Origin and on the Dignity of Procreation" of 1987 was against artificial insemination, in vitro fertilization, and embryo experimentation because they separate the unitive and procreative aspects of marriage. The Catholic tradition finds reproductive technologies immoral because they violate the sanctity of marriage and dignity of individuals, as defined by their premises. The American Fertility Society found the same technologies ethical and in accord with the Constitution. It determined well guided pre-embryo experimentation is "justifiable and, indeed, necessary if the human condition is to be improved" [7]. Of course, pre-implantation diagnosis and selection are objectionable to the Catholic tradition on the same grounds as other reproductive technologies. But, more than that, the totality of Artificial Evolution will most likely be determined unethical because it puts Humanity in charge of a process at the core of the Holiness of Life. Since the belief in Divine Causation and a created universe excludes Darwin's discontinuous non-progressive evolution, its artificial successor will no doubt find even less favor from the conservative–religious right. In general, the grounds for the set of debates will revolve upon creationist issues, the Divine Perfection of Nature and degrees of our intervention that are deemed responsible.

Secondly, there is the fear of risk reaction, "don't disturb the natural order." This response usually states that our species has depleted the ozone, reduced the rain forests, built up nuclear waste, over-populated the globe, etc. and therefore a new evolutionary mode will only deepen our troubles by creating organisms with dangerous and unknown ill-effects. Without supporting the negative impacts of previous interactions or dispensing with respect for unknown factors, it clearly does not follow that eliminating genetic disorders, or the whole field of negative eugenics, is bad or ill advised. Also, it does not follow that transgenetic strains are dangerous. In fact, it does not follow that Artificial Evolution will add to the woe of this world. It argues for the principal of caution [8].

More significant ethical debates center around the "future generations" argument [9]. This argument holds that we should give to a future world and to future generations the same ethical rights we share in the present. The time spans demand equality. The argument originated within the nuclear waste debates of the 1950s. We are asked to imagine a long distance train journey. The staff and drivers, as well as the passengers, will change many times before the end of the trip. Much luggage and cargo comes off and on the train. Then, the query arises, "Should we allow a package of nuclear waste on board"? The immediate passengers, the current generation, that is, will not be affected, but the future ones most likely will be, because the waste will degrade and become more dangerous. We can all agree that such cargo cannot be carried because of the danger to future travelers. So, the nuclear debate was won on this forceful analogy, even though the many plants were built and storage of the waste continues to plauge us.

The same argument is made with a cargo of new genes and the "too dangerous" conclusion is reached. However, many of the new genes actually help current generations and, unlike nuclear contamination, may not be hazardous to future generations. If we allowed the new genes without any regard or oversight there might well be ill effects. If we disallowed the package and its potential altogether, we would be denying change, invention, and the future landscape into which the train will travel. Indeed, we would be refusing natural evolution on the trip as well. Following the analogy, if the new genes are included with deliberation, controlled supervision, and trial-error, then change and the future conditions can be accommodated. If we wait until all aspects of the new technologies are studied or if we wait until a total design for a new future people is reached, we again will be failing to recognize change and the fact that our current perception of our environment, our current view from the train, so to say, will suffice into the future. Actually, the environment is, and will be, changing, and 10 generations hence, our train may well be in space with myriad adaptive zones on asteroid locations, Martian habitats or world space stations. We should accept modification, reassessment, review of the environments and redesign into the future. So, the futures argument concludes that a totalizing plan is not required to permit the "new genes" on board and into our future. It favors openness and deliberate changes without an absolute future vision.

The critical questions are "How much autonomy, how much control will there be for human evolution?" The future generations' argument comes down against Libertarian and Contractarian views, against singular solutions, utopias and central controls and in favor of openness, multiple choices, conscious debate and controls, and an amelioratory attitude toward future worlds. Since Libertarian and Contractarian systems do not fit, perhaps a broad Utilitarian approach is needed.

Another fourth group of ethical arguments focus on "Man playing God." Few of these rise above the fear level of "where will it all end" or "Divine retribution," and usually involve the unidentified assumptions of a cosmic teleology and a continuous stable state for terrestrial species, rather than a discontinuous non-progressive one. However, more complex arguments involve final causes and end-directed issues.

As stated above, Artificial Evolution is a new mode, not the development of another mechanism. Once the genes have been transplanted, substituted or corrected, the DNA-RNA process runs as usual, the genetic codes operate to fulfill their ends, just as in natural evolution. So, consequently, the way in which individuals are directed to their end-goal is the same.

Thus, although the genetic structures have been altered, their fulfillment has not been. The teleonomic directedness of the individuals is unaffected. Teleonomic is distinct from teleological (Pittendrigh, 1958, Mayr, 1988). A teleonomic process or behavior is one that owes its goal directedness to the operation of a program and is different from teleomatic processes in physio-chemistry. Teleonomic behavior or processes are growth, escape actions, migration, food getting, courtship, reproduction, or the passive ones of body systems of the liver, heart or lungs. Individuals grow and mature according to the altered information, the transplanted information of their specific codes, even though it is to ends supra natural or ultra-natural. So, in the terrestrial realm, in the specific individuals, teleonomy is not changed in Artificial Evolution. If anything, the biotechnological processes involved serve to confirm the teleonomic necessity of genetic codes. It would be something if a transgenic goat did NOT yield anticoagulant. But that it does so demonstrates over and over again that the code will run to its end-goal and is so internally directed.

The effect of Artificial Evolution on cosmic teleology is somewhat more complex, for here we ask (1) are we directed from outside to find and to invent Artificial Evolution? or (2) if not, does this then spontaneous invention make us directors of living things?, in a sense final–causers in the terrestrial world?

It might be argued that our new mode of evolution does not affect cosmic teleology because it is a part of it. This argument assumes a priori that there is a final cause to the Universe and that world time has unfolded to let Humankind devise Artificial Evolution as part of a broader historical mandate. Here then the argument would state that Humankind is escaping nature, standing outside of the most basic tenet of nature, evolving out of evolution because it had been ordained to happen in an historical occurrence whose full meaning will be revealed only later, at the end of world time. By this argument, Artificial Evolution, like everything else, is explained by a future revelation that can only be believed. The trouble here is that there is no evidence for this. The argument cannot be disproven, but only accepted. It constitutes a belief.

A more reasonable proposal would be that as an extension of Darwinian biology and genetic biotechnology, Artificial Evolution is simply the invention of Humanity. It seems prudent to regard this new mode of evolution at face value without assumptions—as our invention with an internal source and meaning.

We can now address the second question, "Does Artificial Evolution place us as the final cause here on earth, if not in the Universe?" By evolving out of evolution, by possessing the power to reorganize and direct the genome of species, Humanity itself could arguably be said to be the master of this world, the "what–for," the Mind forecasting and directing nature.

We should recall here the meaning of end-directed: the steps in a chain of events are followed out from a pre-set program or idea. Under the proposal that Humanity has become the "final causer" of living things, we would be going along in a process about which we were completely cognizant, each step known in advance by us. Under the proposition that we are "playing God," we would have to have omniscience, as well as omnipotence.

Artificial Evolution has come about through scientific inquiry. It stems from our understanding of living things and our need to employ the information. Its specific activities have been responses to economic needs, that is, to develop uniform cash crops, like corn or cattle, or to synthesize hormones or enzymes more cheaply. The manipulations of the human genome have been in answer to medical problems, especially inherited disorders. All of these are responses of expediency, rather than premeditation. Furthermore, although there are scenarios that prophesize a grand linkage of the Genome Projects with biotechnology of the next generation, nowhere can the author or authors of such dramas be found. In short, there seems to be no collective strategy for Artificial Evolution. There seems to be no end in our collective mind. There is neither an evil genius nor a benevolent dictator. There is no foresight, nor a collective goal.

It is more fairly said that Artificial Evolution has come about like so many other human inventions, from spontaneity and for expediency. Clearly, then, since its goals, its end, its collective purpose is not demonstrably present to us, we can conclude that through its creation, Humanity has not enthroned itself as the teleological master of nature. We have not the slightest idea of its direction. It will develop haphazardly, at best under ethical review, with conscious action, in fits and bounds and debates.

So, however convenient is the clever phrase, "playing God," we can see that Humanity is "playing Human." The invention that has elevated us beyond nature, that constitutes evolution out of evolution, is without master planning. Artificial Evolution is just happening, in an unforeclosed direction with no end in mind. And, therein lies our burden for ethical, right action.

*   *   *

The Goal Of The Game:
Restructure The Laws Of Nature

The game of the Gene Game turns out to be serious. The definition of human beings, indeed, the existence of new future human beings is involved. Our ability to restructure a central law of nature, evolution itself, and perhaps, then, other laws of physical nature, is the basic goal of the game. The ultimate reason for the game is clear, even if it is being played toward another deeper game. It is our game. And, we must observe the rules carefully, so we can have a great game.

The ethical categories of debate, and, quite possibly, of rule-making will be:

1. *Humanity/Nature or Artificial/Natural.*

2. *Religious concerns, relying upon the dichotomy/unity issue above.*

3. *Fear of risks, to nature, to ourselves, of scientists and by tyrants.*

4. *Future generations.*

5. *Man playing God/Humanity playing Human-- the Teleological question.*

6. *The need for an overall ethical method or system: Bio-Utilitarianism.*

Considering that our chosen ethical system may serve for future transfigurations of other universal Laws of science, our ethics weigh even more heavily upon us. The Utilitarianism of John Stuart Mill that developed in the pre-modern Nineteenth Century stressed the rightness of actions based upon the greatest happiness for all concerned. "The golden rule of Jesus of Nazareth is the complete spirit of utility....to do as you would be done by and to love your neighbor as yourself" [10]. It is the general good and benefit of all that should guide right action. This principle should be the goal of all educated citizens living together, because actions promoting the greatest happiness for all will yield a harmonious society. Mill's utilitarian system was the central liberal doctrine that founded our modern democracies. As we assess the more challenging operations in Artificial Evolution, these Utilitarian principles can be adopted and revised to guide our biological development into the future. We must be lead by an ethical method and Utilitarianism is the best model to allow for flexible development, human wisdom, and common purpose. We need this and can call it Bio-Utilitarianism.

Right action in this new mode of evolution ought to be guided by the openness of information and decisions, by multiple free choices, and by conscious open debates. Centralized control systems and a top-down approach to the specific technologies should be avoided since the processes in human engineering are mainly elective-surgeries, and since they are supervised strenuously by current medical ethics and congressional oversight committees, no additional legislation is required. The respect for life, and its sanctity, is a broadly held concern in the scientific and medical communities. At the same time, the heuristic, therapeutic, social, and even financial benefits of Artificial Evolution ought to allow us to rely upon and trust human wisdom. As the advantages of our new method are examined, a revision of Mill's Utilitarianism, based upon biological grounds, will stand as the guiding ethical system for Artificial Evolution. The same system may well advise us in our greater transformation of other scientific Laws. The utility principle of the "highest social good for all" will remain the best standard to judge our actions and procedures. This cornerstone of liberal democracy will counsel our collective decisions and lead this, our adventure out of evolution.

References:

1. Mayr, E. 1988. <u>Toward a New Philosophy of Biology</u>. Harvard University Press, M.A., p. 318 ff._

2. Hull, R. 1990. <u>Ethical Issues in the New Reproductive Technologies.</u> Wadsworth, CA.

3. Suggested readings on these topics:

    - Dyson, A. and Harris, J. 1994. <u>Ethics and Biotechnology</u>. Routedge, U.K.
    - Hull, R. _____.
    - Smith, G. 1989. <u>The New Biology</u>, Plenum Press, N.Y.
    - Taylor, P. 1986. <u>Respect for Nature</u>, Princeton, University Press, N.J.
    - Von Wartburg, W. and Liew, J. <u>Gene Technology,</u> University Press of America, U.S.A.
    - Walter, L. and Palmer, J., Eds. 1997. <u>Ethics of Human Genetherapy</u>, Oxford. N.Y.
    - Wilmut, I. and Campell, J. 2000. <u>Tae Second Coming</u>. F.S.G., N.Y.

4. Murphy, D. 1999. <u>Transgenetic Animals in Agriculture</u>. CAB Press, CA.

5. Wilmut, I. and Campell, J. 2000. <u>The Second Coming</u>. FSG, N.Y., pp 288 ff.

6. Mayr, Ernst. 1990. <u>Toward a New Philosophy of Biology</u>. Harvard University Press, p. 163 and p. 186._

7. Hull, Richard. 1990. <u>Ethical Issues in the New Reproductive Technologies.</u> Wadsworth, CA., pp. 40-47.

8. Glover, Jonathan. 1984. <u>What Sort of People Should There Be</u>? Pelican. p. 43 and p. 17.

9. Glover, Jonathan. _____ p. 143 ff.

10. Mill, J.S. 1863. <u>Utilitarianism</u>.

SYMBOLS
DICTATES
PLATFORMS
ABSTRACTS

CHAPTER 4

Fascism, Eugenics, Artificial Evolution

Those who fear what they should not fear, and who do not fear what they should fear...go the downward path.

DHAMMAPADA:
THE PATH OF PERFECTION,
1st CENT BC.

The notion of Artificial Evolution is likely to raise in some minds an association with eugneics, and, most pejoratively, with fascism. A brief examination of the eugenics movement and its relationship to fascist policies and programs will help us understand the ethical problems of the "old" eugenics before W.W.II. This, in turn, will reveal how the "new" eugenics since W.W.II is different and, moreover, how the ethical situation surrounding Artificial-Evolution compares to the eugenics movement, old and new.

Eugenics is a word abhorrent to most of the intelligentsia. It conjures up visions of pseudo-science, freak experiments, right-wing politicians and racist attitudes. Yet, new techniques for healing many intractable and congenital diseases, and gene therapy are all praised by the same intelligentsia. This contradiction well demonstrates the division between the "old" and "new" eugenics. In fact, to relieve the stigma of the "old", the word itself is often avoided. For example, we just used "gene therapy" and regularly hear of artificial insemination and in-vitro fertilization without negative reaction, when the practices are often eugenic in purpose.

The eugenic movement originated in the nineteenth century and continued through the twentieth. There are three erroneous assumptions about the early movement: first, that there was a single coherent international approach, second, that the movement was bound up solely with Mendelian rather than Lamarckian genetics and, third, that it was pseudo-scientific. All three have all been debunked by historical analyses of recent years [1,3]. Readers who associate eugenics with fascism and dictatorships will be surprised to learn that the 1924 Eugenic News listed fifteen nations in the International Commission for eugenics, not only the USA, Britain, Germany, and Russia where we imagine the movement took place, but also Switzerland, Sweden, Argentina, and even Cuba. Cooperating nations included Canada, Mexico, Brazil, Venezuela, Columbia, Australia, and New Zealand. New scientific histories show that the movements developed in individual ways in each of these countries and in a fashion peculiar to each culture or more specifically, peculiar to the association or institute in that country. The development was shaped by the nature of the leadership and the particular controlling ideology. In some places, like the USA and Britain, the movement was funded by private philanthropy, whereas in others, such as Germany and Russia, the movement fell under state control. In some countries, research doctors and scientists were in control, while in others, direction came from animal breeders and pediatricians, and in still others, such as the USA, interest in the idea came from social politicians and demagogues. Some nations focused their eugenic programs on "acquired traits", like alcoholism and criminality in Russia, while others, such as Germany and the USA, focused their programs on inherited (Mendelian) characteristics.

For the purposes of this book, we need only to examine several movements, especially in the USA, Germany, and England in order to understand why and how the "old" eugenics became ethically tainted. Then, we can see the situation for the "new", and how each differs from the ethical situation of Artificial Evolution.

Francis Galton, distant cousin of Charles Darwin, first used the term eugenics in 1883. Derived from the Greek, *Eugenes*, for "well-born", it was employed to designate the concern for the growing population and the appearance of "unworthy" characteristics. In Galton's view, the population needed to be "born- well" so the state could flourish. Darwin's theories were beginning to gain wider acceptance and, through Herbert Spencer, their social misinterpretations and mis-appropriations were already underway. The Industrial Revolution had caused numerous changes and disruptions to the society and culture of the time, both in England and Germany.

In Germany, the germanicized term "eugenik" was used only later and for the special purpose of trying to impart a scientific connotation to the movement. The term "rassen hygiene" (race hygiene) was coined in 1895 by Alfred Ploetz. As the movement developed from 1895 to 1933, the central issues concerned the social question of how to care for the people of Germany. Most of the leaders were physicians by training; some, such as Rudin and Lenz who later controlled the Munich chapter of the society for race hygiene and then gained national control under the Nazis, were politically conservative and some were Communists, but most of the leaders, were socialists [8]. The main objective was the betterment of the German people and some solution to the growing problem of a dispossessed underclass. The solution to the social problem was conceived as biological. The hope was to create a healthier, more populous, more productive, and, thus, more powerful nation. The central notion was the rational management of the population as an entity, without much awareness of individuals and their inherent rights [8].

While these biological solutions to social problems and the managerial approach of state policies on the part of the professional class of physicians can be seen as the prelude to latter Nazi state control of "race" issues, the original movement was not racist. The leaders wished to create an abundance of socially productive individuals and to reduce the number of the allegedly unfit. Although some hygienists were Aryan enthusiasts, others in the leadership of the "eugenik" movement condemned aryan ideologies. The German movement flatly rejected the desirability of "Nordic race hygiene." Very little of their literature was devoted to the subject. Ploetz, as the early leader, wrote positively about the Jews in Germany, about their contributions to the intellectual history of the nation and their general social productivity. He favored intermarriage and opposed any notion of separation, or ghettoization [8]. Although it is true that he and the whole movement viewed white Europeans as the model group and mainly dismissed the Africans and Asians as non-participants, it must be admitted that representatives of these groups were few in Europe at the time, and, so, their views may be seen as sins of omission rather than faults of judgment or racist bias. The important fact is that these professional men wished to increase the productive members in their whole society and they sought to do so by pre-selecting for social fitness. The Rassenhygene movement was concerned with class, not race [8]. And, no surprise, it was the class of the founders and leaders that was seen as the supreme model: the urban middle class.

The notion of a bio-medical solution to social welfare issues and the notion of population as a resource for rational management are the critical ideas of the Wilhemine and Weimar years [8]. The eugenicist movement yearned for state control and national policies. The platform included economic incentives for larger families, especially for government employees and teachers, marriage restrictions for the mentally unstable, obligatory health certificates before marriage, restrictions on unfit immigrants, and preservation of the peasant class. They admired the legislative successes in America [4]. But not one eugenic law was passed during the Wilhemine years. Up until the National Socialist takeover in 1933, there was only draft legislation for voluntary sterilization in 1932. And, this draft proposal was based on medical, not racial or social grounds. We shall see how this compares to America.

In the Weimar Republic, 1918-1933, the Rassenhygene movement had a new leader, Fritz Lenz, who did support Aryan ideals and was more conservative politically. This caused a fissure between the Berlin and Munich chapters. The Berlin branch rejected any and all racist views and was more liberal or left-wing as opposed to the Munich group. The organization formed into more local chapters and its intellectual orientation became more strict or rational in its approach to welfare problems. The Berlin chapter adopted the term Eugenik, in emulation of England and America, to indicate a more scientific direction [8].

Oddly enough, the pre-Nazi German eugenicists failed in the implementation of their programs when compared to the successes in the USA, but did form the ground work for national policies of the Nazis, that turned out to be horrific. The managerial logic of manipulating populations and the need for state solutions to welfare issues were adopted by the National Socialists. However, the racist policies of the Nazis did not have the endorsement of the eugenicists. They did not participate in any anti-Semitic legislation [8]. The extensive sterilization, on the order of 300,000, the euthenasia programs, and the Nuremberg laws forbidding racial intermarriage were all Nazi programs, not eugenic ones [8]. This fact is most important as we consider the ethical issues, because it was the fascist takeover of the nation and not the eugenics movement that resulted in autocratic control, racist policies, and all the dreadful acts of the Nazi state.

By comparison, the eugenic movement in America really was racist. It did achieve its legislative goals, ironically enough, by democratic not totalitarian means. But, conversely, it never became institutionalized. And, its racist policies never became the foundation for a totalitarian racist state, which ought to be one indication that eugenics, even racist eugenics, do not lead to totalitarian political states.

The American movement also held a belief in biological solutions to social problems. It too was organized to supposedly upgrade the hereditary quality of the population, to eliminate the undesirable traits by discouraging parenthood of the so-called unfit, by restricting marriage and even by the sterilization and internment of those judged unfit. The program of the movement unfolded between 1905 and 1930. Its agenda was, according to the advocates, the betterment not of the American peoples in general, but specifically the Anglo-Saxon English ethnic stock [5]. The leadership was from this background and upper-middle class. The membership came from breeders, geneticists and scientists influenced by Social Darwinism. And, the leaders, especially Madison Grant, had definite racial and class intentions to their program. [F.] After the influx of new immigration from

Eastern Europe, the Mediterranean, and China in the 1890's, and the migration of African-Americans into the North after the Civil War, the older English American group felt pressured both economically and socially. They saw their numbers diminishing, while the social burden created by the needy was growing. The eugenic movement in America, in distinction to that in Germany, undertook a racist program from the start [5.].

The other major policy of the American movement was the prevention of the "unfit" from childbearing. While there were proposals for all sorts of marriage restrictions, sterilization became the chosen preventative, and non-elective sterilization at that. Between 1907, when Indiana enacted its sterilization laws, and 1931, some 30 states legislated forced sterilization of the mentally unfit. It is estimated that about 20,000 people were forcibly sterilized during this period [5].

Interestingly, during this same pre-Nazi period, Germany did not adopt similar legislation and the largely medical leadership of the German eugenics movement refused to endorse such policies. Germany also failed to restrict or oversee immigration, although, of course, it was a much smaller factor than in America. The U.S. success in legislation was admired by like-minded Germans and occasioned international meetings and considerable communication. [4]. However, after the National Socialist takeover and the establishment of the fascist state, the Americans slowly broke off contact, just as the German eugenicists and physicians separated themselves from the Nazis [8], and that political movement adopted and made horrific the practices begun in America.

The movement in England was very similar to that in Germany, except, rather than a state controlled fascist outcome, there was a democratic and humane one. The very same social forces of industrialization, dislocation, and population growth created a new underclass, the urban poor. This social problem and class became the focus of the movement.

The Eugenics Society was formed in 1907 and was linked to other social organizations of the 19th Century. It was inspired by the writings of Francis Galton. The membership was much like the German movement; it was middle-class and educated. About 20% of members, were social activists, more than in Germany, but the majority were from the scientific and medical community, with some biologists and statisticians. Like Germany, the ideal model type for the improvement of England was their own, the educated urban middle-class.

The social problem of the urban paupers occupied social organizations throughout the 19th Century. The Eugenics Society took a new approach, a scientific and biological one. The attributes attached to the social ills were poverty, alcoholism, venereal disease and low intelligence. The movement, as in Germany, wished to use the biological laws of inheritance to ameliorate these social ills.

From the beginning the eugenicists fastened on a Mendelian biological solution. Since it was believed that the ills of the poor were inherited, rather than social or behavioral, social conditioning as a factor influencing this vast group was not given credence.

Oddly enough, there was very little communication between the English and German movements, despite their similarities and despite the contacts between the US and German societies. The English movement was active within the Commonwealth as a social and scientific society. It never had socio-political leaders or aspirations. Consequently, unlike the German movement, it was not a vessel that could be occupied by political aspirants. In counter-distinction to the German groups, the English society passed quietly into oblivion in the later 1920's and 1930's. [6]

However, the movement in England had the same structure as the other national organizations. They were all top-down. An elite educated group decided to solve social problems by scientific means. The population was viewed as a mass, an abstract unit, to be operated upon. The self-elected society formed an ideal type against which the objectionable social ills were portrayed. A biological-scientific solution, Lamarckian or Mendelian, was forced upon the social problem from above.

In comparing these movements, we can see the main ethical factors in the "old" eugenics. Although the leadership and membership of each movement were different, and although the agendas and their legislative achievements were very different, the ethical problems were similar. The movements viewed the people of their respective nations as a population, as an abstract entity, as a resource to be managed and controlled, rather than as a collection of individuals. Secondly, they conceived the solution to social problems in biological terms. Thirdly, from their inception the ideals of the movements reflected the cast of their leadership. In Germany and England the type grouping was based on class, with that of its founders as the ideal model, while in America the generic grouping was based on race with that of its founders as the ideal. There was little concern about individuals in any movement.

Of course, the notions of class and of race exist as just that, notions. They are abstract entities. This explains their usefulness in politics in general, because they can be employed as concepts, usually for emotional leverage. Individuals remain, however, as material things of great specificity.

This reveals the main ethical crisis of the "old" eugenics: individuals were neither noticed nor honored. After WWII, ethical questions arose from various schools of thought: What about the rights of the people? Who decided? What were the goals of the programs? And, of course, the ethical objections are the same as the list of questions, and the conclusion is the same regardless of the school of ethics: the "old" eugenics was objectionable.

The "new" eugenics developed after W.W.II. A complete understanding of the DNA/RNA functions, an appreciation of sexual survival in the realm of nature and its importance for Darwinian laws, the knowledge of chromosomal structure and the genetic elements, as well as the numerous technical processes involved in gaining this knowledge, all led to the field called genetic engineering. The most common procedures include amniocentesis, artificial insemination, in vitro fertilization, gamete donation, surrogate mothering and the procedures central to Artificial Evolution, that is, trans-genetic operations, nucleus insertions, gene therapy, and cloning. This "new" eugenics reverses the field of the "old", because, as practiced in the USA and Europe, all these procedures are elective medical services offered to individuals. The individuals select the services as a part of their private individual genetic counseling, and, with the exception of surrogate mothering and certain abortions, all are legally approved. Thus, the ethical issues of individuals, their rights and liberties, are reversed, because, while trampled in the "old" eugenics, they are served in the "new". Genetic engineering is elective surgery chosen freely by individuals.

The individual is the basis of current genetic engineering because it is the specific genetic code that defines the individual and, since the services are contracted to correct a family hereditary defect, it is the genes of an individual not a group that are manipulated. But, since individuals are exercising their free choice and desire to remove suffering from a following generation, they are nevertheless affecting the large human gene pool. Even though few would argue with these new eugenic practices, the whole gene pool is something held collectively by all of us, because each of us contributes to the totality. So, individual choices affect and involve the collective whole which belongs to us all. This must be considered our collective heritage. Thus, the ethics of the "new" eugenics engages social and collective considerations on another level and in reverse order from the "old". What were the agreed social goals of the original national movements in the "old" eugenics are now subject to review by the various ethical schools of our own international society.

All this brings up a central problem in the new biotechnologies. Genetic engineering is an open field of individual procedures that is growing. A closed system of ethics can not encompass the new processes. In chapter three, we came down against libertarian, and contractarian, systems, in favor of ongoing debates, a global approach, and an obligation to the rights of future generations and a biological utilitarian approach. From the standpoint of philosophical systematics, all this may seem unsatisfactory, akin to a situational ethics for future generations. Yet, given the developing field of genetic operations, this analysis and approach may well be the most practical and responsible. The highest utility for our species will be our suitability to the evolving technological culture zone within which we live.

In conclusion, the ethical situation of the "old" and "new" eugenics have reversed positions and produced a welter of debates about public policy concerning specific operations. Since Artificial Evolution has not been examined as a process itself, we can only say that it will follow the form of particular biotechnologies. The debates will repeat familiar categories:

1. Religious objections,

2. Fear of risks,

3. Fear about future generations, and

4. Concerns about end goals.

The received ideas about the "old" eugenics include the general fear that eugenics caused or took part in creating fascism. Hopefully, this chapter has clarified some of those early twentieth century relationships. As we have seen, it was an absolute political power in Germany that led to the horrible state policies, not the geneticists or, their movement. To establish a global eugenics against the will of many nations would require a state with total global power. If that power were malicious, its total control would be malicious and its eugenics would be only a small part of an otherwise dreadful world. Thus, it is global control by force that is to be feared and rejected, not a positive eugenics. The conditions for a positive eugenics and the artificial evolution of our genome would hinge on the opposite of absolute control. They would depend on individuals, open debate, multiple and individual choices, and a global outlook across many generations. They would depend on good will and the courage to have faith in human wisdom. A biological utilitarianism will serve us best as the ethical standard for Artificial Evolution.

Thus, the history of the eugenics movement can help us evaluate our species as a candidate for Artificial Evolution. We can imagine some of the problems and situations in an accelerated evolution. Here we confront a really solid, hard ethical question. It is not one that ethicists have approached, or even noticed. It is one all too basic: What species, what kind of creature, would choose to make itself obsolete?

References

1. Adams, M. (1990) <u>Eugenics in the History of Science.</u> In M. Adams (Ed.), <u>The Wellborn Science</u> (p 3.8) NY: Oxford Univ. Press

2. Adams, M. (1990)   <u>Towards a Comparative History of Eugenics</u>. In M. Adams(ed.) <u>The Wellborn Science</u>(p 217-221) NY: Oxford Univ Press.

3. Kelves, D. (1985) <u>In the Name of Eugenics.</u> NY: Knopf

4. Kuhl, Steven. (1994) <u>The Nazi Connection.</u> New York: Oxford Univ. Press

5. Ludmerer, K. (1972)   <u>Genetics and American Society.</u> Baltimore: John Hopkins University Press

6. Mazumdar, p. (1992) <u>Eugenics, Human Genetics and Human Failings</u>. London, Routtedge.

7. Stepan, Nancy (1990). <u>Eugenics in Brazil</u> In Adams (Ed). The wellborn science 110-150

8. Weiss, S. (1990) <u>The Race Hygiene Movement in Germany 1904- 1945</u> In M. Adams (ed.), <u>The Wellborn Science</u> (p.8-50). NY: Oxford Univ. Press

F. The American Breeders Association was founded in 1906; it became the American Genetics Association in 1913, with the Eugenics Record Office established in 1910. The overriding political goal was legislating immigration restrictions. The Immigration Restriction League had been organized in 1894 and its agenda was confirmed and finally enacted through the work and writings of Grant. His book "The Passing of the Great Race", 1916, lamented the dilution of the genetic stock as well as the political status of the Anglo-American people. And, it advocated strict immigration policies. All this led to the fulfillment of the legislative goal of the eugenic movement in the Immigration Restriction Act of 1924 [5].

CHAPTER 5

The Future and Futures

"Meditation brings wisdom; lack of meditation leaves ignorance. Know well what leads you forward and what leads you back, and choose the path of wisdom."

BUDDHA,
5TH CENT BC

Speculations about "the future" vary greatly from our own personal futures. We are confronted with the future of mankind, future society, the technological future, future trading and indeed, the philosophy of the future. The concepts and activity about the future are extensive and, for one very important reason, critical: speculative and imaginative projections in the present seed the future with potential. We need only recall Leonardo da Vinci's 16th Century sketches for a flying machine to understand the provocation that the present can invoke upon the future.

When most of us consider the future, we think of our personal futures. We consider our family plans, our relationships, our careers and investment goals. We think about the months and years ahead, and then, the decade ahead. This personal plane is the most functional notion of "the future".

Philosophers consider the concept of time and the future in the abstract, even though they too have personal timeframes. Other cultures have different ideas about these concepts and express them in different ways as well. Amero-Indian cultures and African cultures represented the concepts in oral-myth traditions, while Western modern and post-modern cultures articulate the concepts ideationally, fictionally, and cinemagraphically. Of course, the ancient Abrahamic religions traditions differ from Buddhist traditions, and both are at odds with current formulations about the future as an idea in itself.

The Hopi culture considered time as a cyclical, sidereal event, because, like many tribal societies, they lived within the natural. The whole of their culture beat with the seasons and their concept of time followed. The past was gone and the future was not yet, even though its season rhythm could be felt. The present was certain, integral, and livable. This same view is held in pop culture.

Some African cultures saw time as a rolling wheel, rather than a repeating cyclical sequence. The GA and Daugme myths show that time was not sensed as a repeating pattern. There was a beginning, a flow of time and a distant future [1]. So, ancient cultures did see time in different ways, and not always as a cyclical system.

Modern and post-modern philosophical beliefs are individual rather than cultural. Considerable debate centers on the concept of the future and how we depict it. Marxists views, for example, are social and material, rather than fluid, unconditional and technocratic. On the other hand, Henri Bergson views the future as open, indefinite and infinite, because he thinks like a creative evolutionist. Martin Heidegger, on the third hand, views the future as limited and finite, held within the state of our actual being, both proposed and conditioned by our past. [2]

So, how does Artificial Evolution fit into future thought? As a process, it is technocratic and sapiens-centric, rather than "natural" and geo-centric. It is our own invention. Consequently, future projections are technocratic, progressive and sapien-scentric. Just as this new mode is imaginative and creative, it yields a future concept that is open and unlimited. Indeed, we will see how this empowering process will propel our species to transform the categories of Natural/Synthetic—Nature/Mankind. Increasingly, our own human nature will harmonize with the broader dynamics of change and adaptation so central to nature.

So, the future concept held here is the same one held by the society that invented Artificial Evolution. The future is open to new definitions; it is technocratic, unlimited, and creative. It is seen and done by us, just as our inventions are.

But, the concept here is more vital, because this particular invention is more stimulating to the future. Artificial Evolution will create a future wherein the human species becomes the central event in our corner of the Universe. Eventually, the very evolution of the Universe may be humanized and directed by us, just as our own evolution now is. Freeman Dyson holds the idea that as we have now discovered and controlled our laws of genetic evolution, we may one day understand the evolution of the physical and chemical Laws and be able to reform and redirect those Laws and, consequently, the structures, the worlds that they build. [3] In this way, it is significant to propose the future as sapiens-centric and creative.

Just as we can see our personal futures, we can project those of our generation and make some predictions about the next decade and about the next 30 years. Practically, in the coming ten years we will gain a good understanding of the human genome and how some of the complex sites interact to produce diseases, behaviours, and characteristics. The techniques for modifying gene locations will progress. Simultaneously, computer-based tracing will enable a quicker and more accessible understanding. The 2000 or so inherited diseases will be spotted and the desire to eliminate them will grow. Parents will increasingly choose which children to birth, according to pre-implantation diagnosis. Selective birthing will become accepted. Some modifications to the human germline may be elected alongside these selective births. Initial successes with chromosome packs may bring breakthrough technologies to greatly enhance disease prevention.

The coming 30 year period will probably see some variation of gene-splicing or gene-transplants with added chromosome packs effecting many plants and animals. Some of this work may be strictly commercial, while some will be medical and some ecological. The future potential for trangenetic work on endangered species can not be under-estimated, because it has yet to be seriously considered. Indeed, our own species may be saved from its dangerous peril by modification to a new nitche in space. Here the social and ethical dimensions will condition events as strongly as the technical breakthroughs. Current work on human longevity could produce dramatic events by 2030, provoking a deeper ethical debate.

Beyond these predictions, a 100-year future can not be suggested, because numerous technological developments will shape the course of events. However, we can follow scientist and humanist Freeman Dyson in his "Time Framing". He describes the century time as tribal or national, the period organized for broad collective achievements. He sees the 1,000-year period as socio-political evolution, the 10,000-year period as a species evolution, the 100,000-year as global and solar evolution, and the 1,000,000-year period as galactic.

"Looking ahead a thousand years, one may predict that the diversity of languages, cultures, and religions will still exist, even if the dominant varieties are different from those that prevail today. The spreading out of human settlements into far distant places will tend to preserve our diversity and at the same time to make our diversity less dangerous. On a time-scale of a thousand years, the genetic differences between human populations may be increased by effects of natural selection or genetic engineering."

"As humanity expands its living space away from the earth, the same processes are likely to occur. Our one species will become many. There is no reason why a variety of intelligent species should not fill a variety of ecological niches in different physical environments, some adapted to heat, others to cold, some to zero gravity, others to strong gravity, some to high pressure, others to living in the vacuum of space."

"On the ten-thousand-year time-scale, qualitative changes dominate quantitative changes. On that time-scale, our values and ideals are totally plastic." [4]

The 100,000-year scale by Dyson is one of spreading across the galaxy in forms unpredictable and unknowable. Yet, the great distances will define a closer tribal form of society.

The 1,000,000 timeframe is intergalactic.

"We today can have no inkling of the nature of our future inventions. The concerns of our descendants a million years in the future would probably be as unintelligible to us today as differential equations or astrophysics would have been to an early hominid roaming the plains of Africa. All that we can say about the future a million years ahead is quantitative rather than qualitative."

"A million years from now, our descendants and their neighbors in other galaxies will perhaps be preparing for the intelligent intervention of life in the evolution of the universe as a whole. That is an adventure whose beginning we can conjecture, but from here it is out of sight." [5]

So, along with Freeman Dyson, the future we see is creative, open-ended, technocratic, and sapiens-centric. Chapter Ten, "Space Culture", will reveal how diverse, elastic, and mobile our futures will be. Indeed, we should always say "Futures" to appreciate the multiplicity of our evolved future. Here we can focus on the responsibility we hold to our immediate 30-year period. The individual processes of Artificial Evolution ought to be available to benefit as many families as possible. We really should and very well can direct ourselves into a larger harmony within the physical and chemical way of nature. We can manage social decency and get closer to the way of changes and adaptability in nature.

Whatever projections that might be imagined can be valued not because they will come to pass but because they might. They are possible. Future images are important not so much to the future, but to the present, because the images germinate possibilities. Speculations that come about from Artificial Evolution should be valued as an impetus into the future.

It is more important to create the avenues of invention than to judge their eventuality. Many of the notions forwarded in this book are done so as potentials, not as soothsayings or predictions. When we imagine the events of Artificial Evolution, it is to create an openness for development. Our time concept of creative openness, limitlessness, and indeterminatness allows and encourages this process. Artificial Evolution is truly an adventure, and we are geographers and voyagers. The greater our imaginative forecasts, the wider and more diverse will be our futures.

References:

1. Kudadjie, J.N. 1996. "Aspects of Ga and Dangme thought about time as contained in their proverbs", in TIME and TEMPORALITY, TIEMERSMA, D. (Ed.), RODOPI, Amsterdam.

2. Herbert, G.S. 1977, Concept of Future in Bergson and Heidegger. Indian Philosophical Quarterly, # 4, p. 597-604.

3. Dyson, Freeman. 1997. Imagined Worlds. Harvard University Press, Cambridge.

4. Dyson, Freeman. 1997. Imagined Worlds. Harvard University Press. Cambridge, p.p 154-5, p. 159.

5. ----------------, Imagined Worlds, p. 167.

CHAPTER 6

Culturing Culture

"Memory is the Mother of all Wisdom."

AESCHYLUS, 480BC

To cultivate the earth by turning the soil, by planting seeds, by nurturing plants, and finally by harvesting, collecting, and crossing them with others, is to culture. We humans are part of, indeed, managers of this natural process. We select artificially and arrange the environment in order to refine a selected strain. So, the wine stalk has grown and developed, and so, too, the hills have been terraced, drained, and whole regions cultured.

The agricultural simile is only part of what culture is, for what we do to refine the earth, we also do to ourselves. We cultivate ourselves and educate our young. We teach them the ways, manners and ideas of our society as well as those of others, so they may grow, indeed, like vines.

What is a culture? A conviviality of rituals and manners, ideas, methods, tools and artifacts changing into something else? In fact, our word culture comes from the Latin word "Cultus". The word itself has been cultured in English. In the fifteenth and sixteenth century it was used with reverential and agricultural meanings, as in has 1420... "In places there thou wilt have the culture, "or in 1538... "The earth... ye brought to marvelous culture." Later in the sixteenth century it was used with educational meanings as in 1651..." The education of children [is] a culture of their mindes". By the nineteenth century it gained the meaning of the development of micro organisms that we use, as in 1884..."When cultures of this bacterium are kept for some time". It also gained the general meaning of civilization, especially with anthropological meanings, as the in 1867..." A language and culture which was wholly alien to them", or, in 1871 the book title..." Primitive Culture."[7] Indeed, what a culture was and how to understand it became a discipline, Anthropology.

Many definitions of culture have been offered throughout the twentieth century, but we should keep foremost in mind the agricultural meanings: to culture is to grow and to nurture. And, as we shall see, like no other time before, humankind is now able to culture culture.

Some writers claim culture to be the distinctive trait of our species, because it is dependent on symbolic thought and language [6,12]. It is certainly a central characteristic of our species. We humans genetically evolved the behavior and capacity to evolve culturally [1, 4]. As an extension of our behavior, the rituals, tools and methods, and artifacts of our myriad cultures are constantly changing, and being refined. Culture is our way of relating to the changing environment. It is a dynamic set of relations, for, of course, genetic developments in our species have permitted cultural developments. Language and communication grew slowly in ancient pre-species; ritual, values and tool making developed more quickly in pre-history, and ever more quickly in the Stone Age.

Other writers, especially biologists see culture in other higher primate species, too, as the totality of behavior patterns that is passed by learning, rather than by genes, from one generation to the next [1]. Thus, culture and its development extend from other organisms and not only culminate in humankind but also, because our capacity is so great, explains our dominion over other organisms and the global environment.

T.S. Huxley holds that much of our behavior is internalized as thought, assumptions, ideas and shared values. "Through human culture, behavior has reached a supra-organismic level" [5]. He delivers a fine definition of culture: it is not a thing, but a process greater than all the artifacts (material results of acts), socifacts (overt acts like rituals and manners), and menifacts (potential acts stemming from assumptions ideas and values). This cultural process is transforming itself in time [5]. Let this be our understanding: culture is a process. It is an extension of our behavior; and our behavior has evolved genetically. It includes artifacts, socifacts, and menifacts. Especially, it includes our tools, which have become an integrated, linked whole.[10]

At the outset of the twenty-first century, world culture is transforming itself by information exchange, thus adopting beliefs and methods one culture to another. All this is carried out by technological developments. The directive and goal of culture remains the same: to create cohesion and to adapt to the environment.

So just as chimpanzees fashion a tool to extract termites from their nests and teach one another the method [1], so, too, we humans culture ourselves to adapt quickly to changing environments through our genetically developed capacities. And, the information is passed down by learning.

Cultural adaptations have some advantages over genetic ones. Once arisen, they can be exchanged much faster, actually in one generation and, secondly, they are cumulative. We humans can learn new processes and values, employ new tools, and fashion new kinds of artifacts within one generation rather than waiting the generations required to effect a genetic change across the whole deme. Cultural adaptations then do not need to be "turned off" as genes must be, or as species often are, to admit change. We remember and can recall, or forget, our inventions. Thus, each generation has the whole cultural past, any part of which can be re-applied very quickly.

We have changed little genetically since the appearance of Cro-Magnon man some 40,000 years ago in Kenya, France and South-East Asia [4]. The changes are almost literally cosmetic. Our cultural growth in the same period is obviously enormous.

These facts have led many to acknowledge that our cultural evolution is faster than our organic evolution. Of course, calculations of speed are treacherous, but examination of the progressive development of stone tools gives an index for and can characterize cultural evolution. It is not constant change, nor is it irregular. Rather, it is increasing constantly. Cultural evolution is exponential [1].

Since learning is the means of cultural evolution and learning is cumulative, it is not so surprising that cultural evolution is exponential compared to the rate of organic evolution. Calculating the speed of genetic evolution, however, is highly complex. First, rates of mutation must be distinguished from the rate of evolution; they are different. Evolutionary change is not steady for a particular species in a particular epoch, nor across epochs, nor is it irregular completely. There seem to be some patterns. But, these general patterns can not fit numerous species or numerous eras. The principle is a steady irregular pace with faster and slower periods. Some eras show new life forms slowly increasing, then peaking in the middle and finally decreasing before the end of the era. However, this can not be applied as a model for numerous eras [8,9]. So, in general, genetic evolution changes at a steady but irregular rate. Our cultural speed is on another order. It is exponential. And, this greater speed does help explain the overwhelming sense that our technological society is indeed speeding up. The changes truly are cumulative and exponentially faster, as compared to the steady, but mostly irregular rate of our genetic evolution. So, it feels like things are speeding up and getting more complex, because they are.

Since the beginning of the Industrial Revolution human societies have been transformed by our tools and technologies: the assembly line, the repeating rifle, the telephone, the combustion engine, telecommunication, atomic power, jet propulsion, the computer and digital processes. We stand at the beginning of an era of global culture that permits multiple local variations, but maintains an overall technological linkage and demeanor. It appears as a whole, almost a force, metaphorically akin to what ancients called "nature". Our Mount Olympus is the internet; the tremendous sky is our satellite network; the distant horizon our urge to innovate. Actually, as we have controlled our environment in order to survive, we have made more and more of those cultured modifications our environment itself. From caves and encampments to cities and states, to a global web-work, we have constructed our own environment. We are now adapting culturally within it.

Take as an example the invention of the telephone. At first, users did not know what to say or how to say it. We learned and taught one another a whole set of rituals and manners, all recently modified by another invention, the answering machine. We do not end phone conversations as we do letters with salutations and identifications, but rather we begin them that way. We say "Hi, how are you, it's me". Whole sub-cultures and sets of inventions stem from the telephone. We adapted to this cultural adaptation that allowed us to tele-communicate.

The overriding principle of the Late Industrial-Digital Age has been the shrinking of the globe and the quickening of communication in the process of unification. Our geometry is flat; the third dimension has become a point on the screen [12]. Our physical location has been reduced to the time of a signal across the global Netscape. Our methods, our manners, tools and even new inventions have much more to do with the cultural than natural environment. This is pointedly so in the Digital Age, because so much change has occurred by making algorithmic relations by computer of older analog methods. Pay checks are computed rather than typed; music is compacted to digits rather than recorded; images are digitally designed rather than drawn or photographed. We are adapting culturally at an exponential rate to our cultured adaptations, which have largely become our environment.

Our human capacity for culture has propelled us to become the dominant species and this same capacity will be the condition for the future. Considering that our cultural, or psycho-social, evolution is so much faster than our organic evolution and that so much of the change occurs within and according to the cultural environment, and considering that we have developed culturally the mode of Artificial Evolution, we are posed with an interesting dilemma. Really, it is more than that: we stand before an enormous potentiality.

Our genetic evolution has been steady and irregularly paced with little environmental pressure in the recent past, while our cultural evolution is exponential. We now have the power to change the speed and character of our genetic evolution to match that of our cultural evolution. We can redirect and redesign our species. Since Artificial Evolution is a cultural development and since Humankind now resides mainly within a cultural environment, it does make some abstract sense that our organic evolution ought to be cultivated, so as to match rates.

This then is what the circular phrase of our chapter heading, Culturing Culture, would mean. If we could undertake our own genetic evolution, we could enlarge our capacity for cultural development and adapt more completely to our new technological environment. If we recall the underlying agricultural meanings of culture, we would be growing and nurturing ourselves so as to enhance our cultural abilities. And, of course, accelerating our cultural achievements would inspire further genetic change. This linguistic circularity, culturing culture, indicates something much more dramatic, akin to a spiraling effect of exponential change for humanity and our world.

Our inventions and cultures would change dramatically faster. We would too. This would entail improving our capacities for communication and systematics, expanding our intelligences, enlarging our receptivity to information, and enhancing our drives for creativity. All this would be redesigning or replacing our species, just as natural evolution has in the past and would in the future if there were survival pressures stemming from our new technological environment. Since by general agreement there is little pressure for natural selection [1, 2, 3, 4, 5, 9] Artificial Evolution could take us forward, according to its characteristics. The changes could be fast, as fast as our future cultural changes, and immediate, as quick as each generation. Furthermore, they could be specific, chosen by us, and not random. And, of course, we could adjust the pace at any point. We could begin an evolutionary branch uncommon to Homo sapiens; we could descend uncommonly, artificially, indeed extraordinarily. Clearly, this would be a worldwide undertaking and would involve an enormous, some would argue, horrific paradigm shift in our awareness of ourselves and our planet.

Naturally these changes involve highly abstract qualities located on numerous gene sites; remodeling of this kind is far beyond the capacities of contemporary engineering. Curiously, the more specific becomes the image of a transformed humanity, the more preposterous becomes the general notion. Perhaps, this is because our self-identity is so strong that any futuristic depiction becomes somewhat grotesque. Of course, if we adhere to the "future generations argument", then we must develop little step by little step, according to notions responsible to generations far ahead. We would not have to agree upon a future human species before we set upon small immediate changes. So, reforming the species and controlling its future evolution is the proposition of "Culturing Culture". And, the ethical issue is monumental.

But, let your mind wonder on some of the possibilities. Presuming the grand assumption that we could acquire the knowledge to affect our genome, imagine a human hand with seven instead of five on digits; imagine a mind with more endorphins and faster synapses; imagine an expanded sensory apparatus multiplying our audio and visual capacities; imagine being able to handle 25 things at once instead of 5 or 6.

This spiral effect of culturing our cultures would be a lift-off for the whole human race. We would shift out of the natural process by an invention of our own cultivation. By understanding, and then commanding, the natural law of Evolution, we have devised a way to spin our world even faster. As the speed of genetic evolution matches that of cultural evolution, of course, the global change becomes exponential too.

Yet, it lacks subtlety to view this transformation as a contest of development, or as a speed trial. By evolving artificially, we would equalize characteristics more complex than speed. We would be able to adjust our genetic natures to our cultural natures. For example, musical abilities might be enormously enhanced and the hand with seven, rather than five, digits, would be especially useful. It would be as though certain design perspectives were isolated and specified, rather than randomly given, as musical genius now is. We would be cultivating a unique aspect of our human nature to fit a specific task, such as musician. We do select children and train them to this same career today by education and musical discipline. As it is, we accept the genetic formulation, but, when we choose to cultivate them, we would create not only mission directed people, but also whole new musical worlds. These new musical dimensions and experiences would call forth the need for more refined abilities, which would in turn lead to more musical styles. While this does seem extraordinary and speculative, it is important to remember that we express the same process currently with given traits in children. We all attend the opera or symphony and applaud the success of musicians who have trained and specialized their given talents, so, developing them genetically and enabling the whole process is not so largely different.

As a more extended example, Humans of distinct stature and appearance, with capacities to tolerate much more carbon dioxide or unusual atmospheres or gravities or temperatures might evolve for specific missions. Indeed, the colonization Mars would be a perfect current application. We conceive of transporting whole Earth bio-systems up onto Mars or building protective enclosures for earthly humans, rather than the other way around. Homo sapiens developed for grassland zones eons back, about 120,000 years ago. All organisms evolve to a particular niche and our current conceptual framework is to build our earthly one out into the Solar System, when we should be evolving the organism to match the environments that exist out there. By this redirected thinking, mission specific humans would sent forth on unique sorties. For Mars, the human capacities would be extreme because the job is extreme. Yet, this example reveals the more subtle aspects of culturing ourselves to our cultural needs. If we imagine the inter-relationship of urge and response, it is akin to psyche and soma. The cultural requirements call forth a genetic type; the new human type then creates a cultural form that engenders another human development. This is the elastic call and response that, in the abstract, exists in natural evolution where changing zones call forth new life forms, where changing environments demand adaptation through sexual selection and survival. Here it is all happening in the human realm. It is artificial and beyond, but parallel to, nature. Our cultivation of culture then implies the cultivation of the Solar System.

The analogy of spiraling does expose this aspect of the transformation, because a spiral widens and expands as it grows, in contrast to a straight or curved line. Our genetic change over the past 40,000 to 75,000 years is analogous to a flat line, while the notion of "culturing culture" compares to a spiral. This opening and expanding image helps conceptualize the character of the change just now beginning for our species.

It indicates the direction out beyond our globe to the Solar System and shows the limitlessness and openness of our futures. By co-ordinating the speed and character of our genetic and cultural change, we will be enabling ourselves to inhabit the wider environmental zone of our Solar System.

Well, "Why would we want to go any faster anyway?" is one of the myriad questions that arise. Numerous writers conclude our species is slowing genetically, because the environment has been controlled, and thus, it doesn't place any force on us to adapt [1,3,4,5,9]. But, more than that, mammalian evolution in general has been slowing and given past patterns for other life forms, might indicate the possibility of another life revolution. "Mammalian evolutionary activity has started to decline in the last 10,000,000 years or so. This might suggest the interesting possibility, on the basis of the regularities of past patterns, that the arrival of another group is due" [9.p.112]. Perhaps, Artificial Evolution is just exactly this change. And, the other group is our own species modified for the mission.

Since we are at the beginning of the era of space exploration, evolving ourselves into the apparatus and environment of our own making should facilitate the quest into space. We might be changing ourselves to fit an environment where we could be much more effective, if our organic nature were quite different. Future humans may be artificially evolved to tolerate much colder or hotter climates, or much lower gravity forces. The body processes may be much slower. They may have radio reception/transmission capacity and infrared vision. Indeed, they may not be carbon based. Most certainly, they will be multi-formed.

In some mysterious way, the expansion of our environment into space may explain our new mode of Artificial Evolution. Perhaps, an extraordinary new environment in space may be requiring genetic adaptations that can only be managed artificially. But, in any case, the answer to the question above- "Why do it?" - involves social and ethical discussions of the broadest sort. Since we circumnavigated and mapped the globe in the Age of Discovery, our species has not faced an enormous expansion and the frontier has been largely inward, psychological and earthward. The Space Era ahead may require, and indeed, may be calling us to modify our genome to match the culturally devised environments of a vastly larger habitat.

Now, of course, this prospect ought to set off an ethical alarm in the minds of most of you readers. Even if we know "how" and "why", we would have to know "in what way". To culture culture, to take up the restructure of our genome and gear it to cultural developments, entails a positive, and positivist, bio-technology. So, we must return to ethics and rely upon the Utilitarian System discussed. The arguments will unfold as we predict in chapter three. But finally, they will rest on a trait borne of ancient organic evolution and refined by the cultivation of millennia: **human wisdom.**

References

1. Campbell, B.(1966), <u>Human Evolution</u>, Chicago: Aldine Co.

2. Darwin, Sir C. (1960), "Can Man control His Numbers," in Tax S., <u>Evolution After Darwin</u> (II, 468 - p.474). Chicago, University of Chicago.

3. de Nouy, Lecomte, (1947). <u>Human Destiny</u>, NY. Longmans Green, Co.

4. Dobzansky, T. (1963), <u>Mankind Evolving</u>, New Haven: Yale University   Press.

5. Huxley, J.S. (1958). "<u>Cultural Process and Evolution,</u>" in Roe and Simpson, <u>Behavior and Evolution</u>, (P. 432-54), Yale University Press, New Haven, CN

6. Kroeber, A.L. (1952). "Culture: Critical Review of Concepts and Definitions", in <u>Peabody Mus.#47</u>, Boston: Peabody Museum.

7. Oxford English Dictionary, Second Edition. (1989). Oxford, England: Claredon Press.

8. Simpson, G. (1942). <u>Tempo and Mode in Evolution</u>, New York, Columbia University Press.

9. Simpson G. (1967). <u>Meaning of Evolution</u>, New Haven: Yale University Press.

10. Tax, S. (1960). <u>Evolution After Darwin</u>, Chicago, University of Chicago.

11. Virilio, P. (1991). <u>The Lost Dimension</u>, New York, Semiotext.

12. White, L.A. (1959). "<u>Concept of Culture</u>". American Anthropologist, 61.

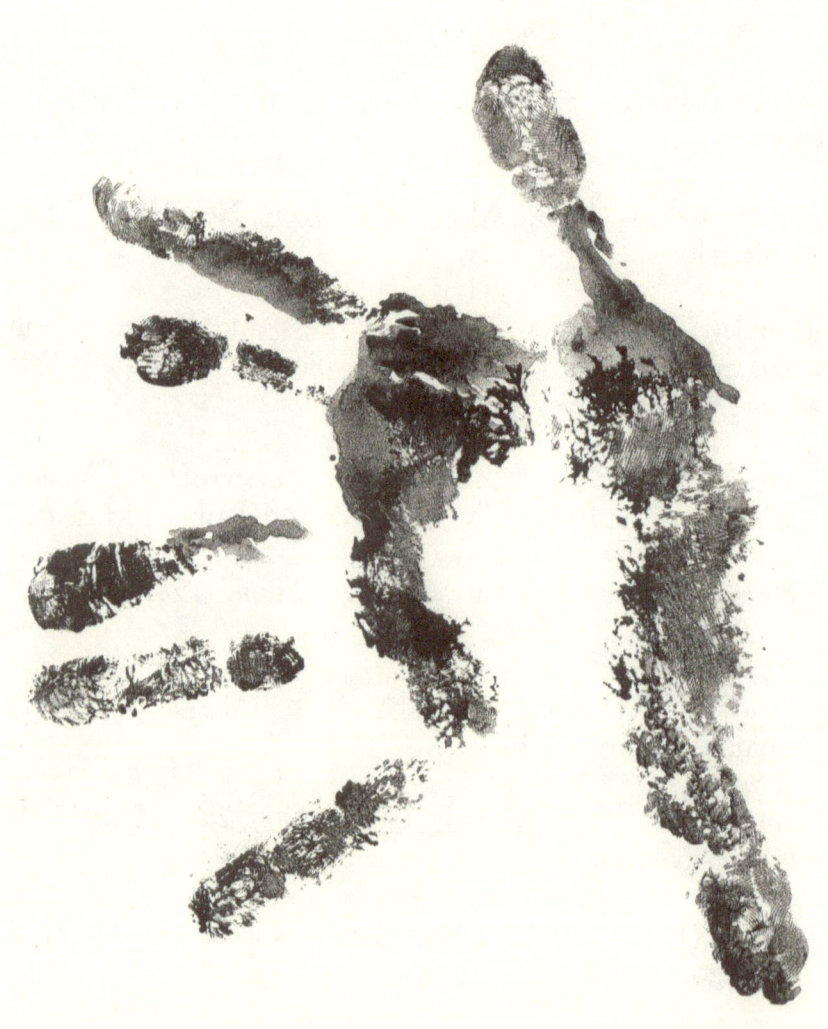

CHAPTER 7

Homo Sapiens Novus

Creatures of a day,
What is a man?
What is he not?
Mankind is a dream of a shadow.
But, when a god-given brightness comes, a radiant
light rests on men, and a gentle life.

PINDAR, ODES, BOOK 8

Each member of every species competes to survive and reproduce. The urge is to continue, and in effect, to adapt. So, what species would choose to halt itself, to discontinue, and to replace itself with another form? What species would transform itself?

Our new mode of evolution results in new living things, such as transgenetic animals, plants and cloned beings that exist in a variable evolutionary mode conditioned by humankind. By definition, they are not species. As they came from Artificial Evolution and live in its laws, the way that they came together is different. While they can and do reproduce sexually, often it is asexually. And, they are not selected naturally to reproduce. Most importantly, the gene pool is never stabilized, but rather is always mobile; it is created to change immediately. The group members are not isolated genetically; they exist in a fluid state of immediate change governed by humankind. This could be called a sub-class, perhaps sub-species; "Dolly" could be described as a super-organism. If a whole herd were maintained as a continuous clone, this would be yet another type of organism. If a particular sub-class had numerous genetic additions and deletions, it might be termed sub-species. Furthermore, numerous transgenes might isolate a group and result in a new species. The questions for taxologists are as perplexing and complicated as the life forms. New categories are required. But clearly, we should consider these groups unique. The concept of species is different. To indicate this distinction and yet avoid a new word for these organisms, let us bracket species: {species}.

While sex selection by pre-implantation diagnosis or, more often, by abortion is a significant ethical and social problem [1.], the general prospect of selected characteristics in a positive program creating new humans is not much of an ethical dilemma, because, of course, it is not much in practice. However, it is not that very long before germ line engineering, the main method

100

of Artificial Evolution will be safe and available. The recent completion of the genome project by the NIH and Celera Corporation has provided a chart of the 33,000 active genes in the human genome; the interaction of sites and the complexity of expression will require many more years of study. But, within a decade, our understanding will be extensive, and our ability to represent the complexity by digital means will lead to broader and simpler presentations. The relation between a genotype, the individual genetic structure, and phenotype, the real living person, will become readily understood and normative.

So, let's consider a really extensive positive program. Very conjecturally, let's suppose we understand the whole genome and how its locations interact to produce a phenotype, an actual person. Let's suppose we can technically achieve a restructuring of the genome involving new loci by replacing and adding genes and artificial chromosomes that would result in a major set of changes: a larger cranium with more endorphins and synapses, a more active memory, an expanded sensory apparatus, a body frame more suited to our cultured environment, and an extended life span. Such an effort would produce a DNA structure so different that it could not match in combination with our original one. In other words, we could not interbreed with the humans of a really positive program. We would have created a new {species}. Say, Homo sapiens novus.

If we accept the theory of Artificial Evolution and a new line in evolutionary history, if we accept that mammalian development in general has slowed over the past 10,000,000 years (2.), if we accept that our species has conquered the physical environment in general and is now developing in a cultural environment and, thus, like other mammalian species, receives little pressure to evolve organically in order to survive, if we can agree that sapiens is little changed in the past 10,000 or 30,000 years (2), if we accept that cultural evolution changes exponentially while natural organic evolution changes gradually and periodically [3], if we accept that our cultural evolution presents a new environment wherein new facilities would be more desirable, and if we can accept the more conjectural notion that organic evolution could match the exponential rate of our culture change, then we are prepared to agree to a positive program. This, of course, endangers our species and is the chasm of the logic: why would we replace ourselves?

Artificial Evolution is real. The procedures we have discussed are on-going. The laws explain the new processes in the simplest fashion; they are correct. And, furthermore, with each passing year, they become more practiced. New lines of evolution are being created. We will create many new sub-species. We will develop new forms of endangered species, the un-dangered. The more positivist is our program, and the more utilitarian is our ethics, then the more extensive will be the humanization of the globe. Imagine the taxonomic charts in the natural history museums that show the linear branching of species. They all descend to the year AD 2000 and then the various branches begin to torque and cross and swirl. By 2,500 spirals will gyrate in 3 dimensional patterns of accelerated speed. AE is indeed a new law, and it will yield extraordinary speciation. Our own human development must be chosen with care, but also, with imagination and vision.

Is it true that mammalian development has slowed over the past 10 Million years? Has the pace of speciation slowed, because the zones have been so successfully occupied? Yes, according to evolutionary biologists [chapter 6]. There is evidence that such slowing trends presage the advent of new life forms [3]. If mammalian speciation is slowed then a long period of stabilization has been underway. The decimation of natural habitats and the addition of toxics into life zones are causing a mass extinction of species. The natural evolutionary process of reinvention will not be able to react quickly enough to this crisis. New zones in and around human habitation will open up for the remaining species. While natural evolution might manage to deal slowly with natural geo-physical changes, these new conditions are created by human culture. Thus, human processes must solve those problems. In addition to all the other ecological repairs we must undertake, we must also pick up the evolutionary pace for the many species that are being affected and destroyed.

Obviously, not every genome can be examined and modified, but then, argumentatively, not every life zone is being reduced and polluted. We can and will have to choose to intervene in the evolution of numerous endangered species, or else, we will have to turn them into living museum pieces. The various amphibians that are in distress might be model cases for intervention. Success with some endangered species might show the way for others.

We thrive within the total living eco-system. All species are connected within and throughout their zones. This includes our species especially, because we have become the dominant one that is manipulating the global environment. Therefore, we should subject ourselves to changes along with others. Either we must change or perish, or build a bubble around our current state. Our species branched off perhaps 120,000 years ago, and obviously, the evolutionary change is out of our collective mind. We cannot recall the birth of our species, because it long preceded our consciousness as Cro-Magnons. But, Artificial Evolution does provide a way to quick-start our change in the global conditions that we have created. All species are effected, but we are most of all, because we depend upon so many others and are the most vulnerable.

This is especially true because over the millennia of recent civilization, we humans have learned to cultivate our habitats and protect ourselves from the environment. We have settled in city-states, we have controlled our water and food requirements, and insured our survival. The members of our species from 10,000 years ago, if not 30,000 years ago, could be introduced into our current culture as adopted children and we could not recognize them as different. We have stabilized our growth as a species by controlling our physical zone so there is little pressure to adapt. As our cultural evolution has changed, we have remained genetically stable. While this has not been disproportionate in the last 200 years of industrialization, it is becoming critical in the period of global change, digitalization, and space exploration. Our expotential cultural change is starting to dramatically outstrip and outpace our genetic stasis.

Of course, we can choose the absolute preservation and conservation of our current species; we have the same option with all other species. However, as the gap widens, our limits will become ever more obvious. We are now abiding within our cultural habitat and soon our genetic restrictions will inhibit our cultural invention, because we will not be able to use the very systems we invent and they will begin to appear unlivable or unsuitable, just like a new environment. In fact, the whole cultural zone of space exploration does, and will, appear uninhabitable until we decide to evolve ourselves to fit into it.

Many aspects of our genetic make-up could better suit our cultural habitat even today, if we decided to change. One of the vexing questions in the debate about a positive redesign of our species concerns making our own genetic development a constituent of cultural invention. The same rate of change could exist in our organic nature as in our mechanical world. Indeed, we could design both in synch to specific purposes. Thus, the exponential change in culture would be reflected in our genome. This is not only a positive program, it is a positivist one. However, it does open a philosophical chasm.

It is a perturbing ethical problem. Of course, on a simple level a screeching "out of the question" can be heard, or, "We are not going to make ourselves an endangered and then finally extinct species". But, it does get more complicated and the question of what species would make itself obsolete becomes more probing. A truly positive program involves the creation of a new species, or group of sub-species, and presents not so much an ethical problem as an ethical abyss.

The canyon of the ethical question spans many generations. Certainly, a new {species} or group of {species} would take ten generations, perhaps 300 years to establish, even though small groups might be created for specific environments within 100 years. Projecting our ethical concerns upon them and imagining their responses, assumes the {species} to be like us, when clearly they were created to think differently.

The ethical situation is difficult to analyze because we have to ask questions not only of ourselves in this generation but also of another generation in another {species}. Would the greatest good even appeal to another {species}? Would liberty be a desirable choice to another human {species}? Does not the notion of another {species} smash the concept of an ethical contract? All of these, and many more questions, falter upon the fact that if we created a new group of humans, we could not know their minds because, obviously, we structured them differently. Yet, there may be some ways to come to terms with this situation. Let us set aside the fearful reactions, such as a group of super humans enslaving or cannibalizing us, such as mass killings to make way for novus sapiens, etc., and attempt a utilitarian evaluation. How might a new line of humans serve the greatest good?

If we can define the common good as the harmony and utility of the whole group within the given environment, and if we agree that our cultural evolution is creating an environment that we can not keep up with and, in fact, could be better adapted to, then we can see that the greatest good would be served by developing ourselves genetically to fit better into our new and ever changing cultural environment. Oddly, quizzically even, the greatest good would be best served by transforming ourselves.

If we weigh our responsibilities to future generations, say, ten generations hence, which would be about the year 2,300, we can postulate our cultural growth at exponential rates and wonder about enduring in such a changed world. It would seem responsible to give the most secure and harmonious life to future beings. The future generation argument asks what responsibility we have to future people, and if we should allow step by step changes in the genome without a complete blueprint for new beings. Clearly, we cannot establish a contract with the future, but we still must honor and provide for our offspring. If we deny any and all changes, we ignore change, invention, human nature and natural evolution itself. If we demand a total design, we consign ourselves to our current views, values, and interpretations of the possible distant environments, which are all limiting. Thus, a middle way of deliberate conscious steps in Artificial Evolution seems the most prudent. So, here too, curiously, the situation argues for the enhanced development of our species, so that a new group can live the most fully within their environment, even if it is a new {species}.

So while the thought of extinction is repugnant, there are some solid ethical reasons for choosing a positive program and a new {species}. A share of the problem in analyzing the issue is that it is our species, ourselves, in question. If we considered another species, say the endangered condor, and proposed altering the genome to the degree that it became a new form of condor, one more suited to the humanized environment, and then put the utilitarian and future generation questions to the test, most of us would agree that the condor would be better off supplanted by a new form. Of course, since the condor is already endangered, its survival is the crucial factor in winning the agreement, and most of those who hesitate have to give up the notion of the maintaining the 20th century environment ten generations hence, in the 24th century, in order to consent. So, even though comparing species is not completely fair, it does reveal our own protective attitude toward ourselves. And the comparison does reveal the fundamental truth that every species will have to adapt to the humanized global environment or become extinct. In the case of our own species, that means the global cultural environment.

If the new species proposal were agreed upon, how would it come to pass? Would there be normal births? Would current species members be allowed children? Would there be a global health organization? If we can put aside fears of a totalitarian state, mass sterilization, camps, etc., we can still question the actual function of such a program. It would seem that much of the ethical debate and even indignation concerning the proposal of a new species rests upon the very human issues in implementation. Just how would a new species be brought forth and welcomed?

Since utilitarian and future generation arguments favor a positive program, the functional questions will remain the most open of all those proposed. Of course, there will be no easy answers. But the most likely scenario is individual couples exercising their reproductive rights and choosing to enhance the character of their children. Step by step, the new {species} would take shape, based upon diligent practice and conscious design. So, far from a vast social program or theoretical impasse, the process will most probably be intimate and humane.

Lee Silver has described a couple in the year 2050. The mother and father view a computer screen for images of their possible children. They select from several hundred. The immediate physical appearances are projected, such as sex, height, weight, hair color, etc. Then, there are the lists of congenital genetic disorders, predispositions to complex and infectious diseases, physiological characteristics, and personality and cognitive abilities.

The mother has had a small part of her ovary removed to provide thousands of eggs that have been nurtured to readiness for fertilization. Several hundred have been fertilized and incubated. The pre-embryo masses have provided individual nuclei for pre-implantation diagnosis. The gene locations and their interactions have been analyzed and digitally presented on screen. Pre-natal parenting begins much earlier with selecting the child to nurture.

The couple rejects the genetically disordered combinations; they avoid those predisposed to serious diseases. The physiological combinations are varied and the choices are personal. Temperment and cognitive abilities are refined judgments, too. Finally, they select a girl with musical abilities that reflects both parents, realizing that there is no perfect person, not even for them [4].

The multiplicity of offspring from these parents and the complexity of their selection indicates that such a pre-natal method would not be typecast. Indeed, it would be an elective surgery, conscious, directed to good health and based upon intimate preferences. Far from any state mandate, it would be personal and humane. This family practice is likely, because, since parents already spend considerable time and expense rearing their children and pre-natal care gets ever more extensive, they will want to select which of their off-spring to rear.

This is the most likely avenue toward an enhanced new {species}. The next step would be exchanging genes from one nucleus to another. Parents might select every aspect and then change the sex; or, they might combine traits to design just what they want. A next step would be genes from another family, such as low cholesterol traits lacking in their line. Finally, genes from other organisms would be added to yield special characteristics. Attributes from other mammals would be tried, such as those that prevent HIV-AIDS from Baboons, or sonic-echo genes from Bats. Slowly, the human family would expand.

This step-by-step transformation might occur in another way as well, based upon a human process as fundamental as family. Our higher education system produces specialized individuals through cultural learning and refinement. They are the product of not only their family background, but also their institutional training and formation. This is true for professions in law, medicine, aerospace, and even gymnastics. The individuals are formed intellectually and even physically by their institutions; actually, most of us can easily recognize people from the conformation of their profession; people look like what they do. It is but a small leap to imagine the same institutions forming their candidates genetically. At first, it would be in the specialized fields of rare service, such as the space programs or perhaps language professions. Imagine the NIH and NASA cooperating on child rearing. Although an unlikely proposal, parents might choose to join such a program. Over a century, a special service approach would yield a range of new humans. Quite likely, this would in turn expand the perspectives in parenting as well.

Both family and an institution-based development rely upon age-old human practices. Child rearing, nurturing, and education are fundamental to our way of life. The process extends and perfects the very human practices most common in society. Thus, the enhancement and redesign of our genetic nature would follow normative and humane paths. New {species} would evolve from both our natural genetic and cultural heritage.

Another ethical question reveals an even larger responsibility. If we agree with Darwin's view that natural evolution is non-progressive, discontinuous, and beyond values, then creating a new human species is an ethical act on another order. When a species becomes extinct, another takes its place, most of them in the changed environment that caused the extinction. A new species is not absolutely better or more advanced, but only relatively better suited to the new environment in which it is surviving. The environment of the planet is not better or worse in any epoch, just as the current aquatic environment is not better or worse than the terrestrial one. This is to say that natural evolution has been non-progressive for the whole run of life, despite the flattering notion that our species is at its pinnacle. But, Artificial Evolution would change that, for it is controlled to be progressive. The current transgenetic plants and animals have been created for specific ends that their natural cousins cannot achieve. All of them are better from a technocratic and homocentric point of view. They are progressive and have value to us. In this sense, Artificial Evolution would differ from natural evolution. It would be value laden, progressive and continuous.

Artificial Evolution would bring progression and human value to evolution and to a new {species}. The historical and ethical import cannot be underestimated. This would be true of the whole process and it's created {species}. So, the new humans would be better able to serve human needs, rather than merely adapted to the grassland environment eons ago. Thus, from an ethical point of view, humans could be said to benefit from the creation of value and progress in evolutionary history, in spite of the fact that they would be eliminating themselves, because they would be better at being of and for human needs, better in the human cultural environment, and presumably, as created, better for artificial environments of the coming space era. Thus, the passing and replacement of our own species might be acceptable, if a credible new form of greater suitability were artificially evolved.

To consider and imagine this situation as a "replacement" or "passing", of course, commits an error central to hominid anthropology in the 20th Century [5]. We assume the current single species of man to be typical of all times, past, present and future. Until recently, hominid evolution had been considered as a lineal development, one species after another, rather than a taxic complexity, with numerous species co-inhabiting and competing. There is now "increasing documentation of substantial bushiness through out most of hominid history" [5.]. Current postulations have four hominid species co-existing as little as 30,000 to 40,000 years ago: H. neandethalensis, H. erectus, H. floresiensis and H. sapiens. With H. sapiens being the only current species, "our modern world represents an oddity rather than a generality." "The general view is that during most of human history, a multiplicity of human species lived at the same." [5.]

This realization of our past taxic complexity and co-habitation with other human species was brought dramatically forth in the discovery in October, 2004, of the "Hobbits" of Flores, Homo floresiensis. [8.] [9.] The general agreement is that the "small people" arrived on the Island some 84,000 years ago, evolved from H. erectus or possibly by fast paced evolution in the confined island habitat, which would follow the Mayr-Gould Punctuated Evolution Theory. Certainly, the species existed until 12,000 years ago when a volcanic disaster buried much of Flores in ash. Since modern humans migrated into

Australia and Borneo about 40,000 years ago, there was a 25,000 year period of possible interaction. Island mythology will perhaps have some bases in fact as the discovery process continues. In any case, the "Hobbits" have made the evolution of humans and our co-existence obvious.

"The biological history of hominids resembles that of most other successful animal families. It is marked by diversity rather than by linear progression.... It is evident that our species, far from being the pinnacle of the hominid evolutionary tree, is simply one more of its many terminal twigs". [6] If we believe that our current single status is unusual, then it may provide further proof of the stasis in our genetic development. As a single species we can only change in a linear way, by transformation. Favorable traits will have to arise within our same species and compete in a changing environment. This is very different than a variety of species in a broad nitche with a multiplicity of traits. Clearly, there is not another species today, unless a surviving group of Homo floresiensis does come out of a hidden cave. It is fairly obvious that our single species does not have unusual traits, such as the capacity for much more carbon dioxide or to tolerate much lower body-core temperatures. In many ways, then, we can see the stasis of our genome in our isolation. It is also our vulnerability.

In a radically changing environment, our species would have to generate traits within itself. There is no group of species that might have the necessary traits to survive. Since there is only one species, the origin of a key attribute is less likely than if there are a number of species. Our ability to change and to adapt to an extraordinary situation is limited and even dangerous.

In <u>Animal Species and Evolution</u>, Ernst Mayr wrote:

> It is the very process of creating so many species which leads to evolutionary progress. Species, in the sense of evolution, are quite comparable to mutations. They are a necessity, even though only one of many mutations leads to a significant improvement of the genotype... It appears then that a prodigious multiplication of species is a prerequisite for evolutionary progress... Without speciation, there would be no diversification of the organic world, no adaptive radiation, and very little evolutionary progress. The species, then, is the keystone of evolution. [7.]

There is both archeological and genetic-dating evidence that our species shared the world in the past with other hominid species, but there is also proof on principle. An isolated single species cannot adapt as well; it is not as enabled as a multiplicity of species. This argues for the taxic complexity in our past, but it also argues for a multiplicity of human species in future. If there were a single hominid species from the outset that transformed linearly, Homo sapiens probably would never have arisen. The single ancient species probably would have failed. Bushiness is better, and it is more historic.

So, although it is commonplace to think sequentially and linearly about a created new species following upon and replacing our own, it is truly more historic and normative to imagine another species co-existing with our own. Our co-inhabitor, or co-inhabitors, would be crafted by us, of course, but still, the global or global/space environment would be alike numerous times past, and indeed, typical of past hominid evolution.

If we accept the theory that an abundance of human species existed simultaneously in the past, we have overcome the fallacy of projecting our contemporary sense of things backwards. This then may help us project forward and conceive a multiplicity of human {species} in the future. It may salve the wound, even the horror of, transforming ourselves into other forms. Moreover, it shows that a complexity of human {species} will provide higher security of survival and adaptability into the future. Our survival as a species may be aided therefore by new {species}; indeed, it may be our guarantee of adaption. Pure survival by new {species} may be the override, or trump card, to any and all ethical hesitations, because without some form of humanity, ethics becomes mute. A variable group of {species} that can quickly adapt into other forms will be the keystone to our future evolution.

This, then, is the game of the Gene Game – the happiness of survival and the legislation of new Scientific Laws to transform and, indeed, humanize the solar system. The game is to bring meaning and progress into evolutionary time and thereby humanize the natural laws of the Universe.

What species would transform itself? It turns out it would be the wise species. It would be the species capable of understanding the common good, capable of appreciating its own evolution, and wanting to add value and progress to its evolutionary history. It would be the one that could craft a tool out of a basic law in nature and use it to shape a new creature for another habitat. It would be the strategic one. It would be the species capable of seeing a new horizon and seeking it; the species capable of adjudicating and correcting its own actions. It would be the one capable of responsibility for itself, the wise one, Homo sapiens.

References

1. Kumar, D. (1987). Should One Be Free To Choose The Sex OF One's Child? In R. Chadwick (ED), <u>Ethics, Reproduction, and Genetic Control</u>. (P 172-182) NY: Croom Helm.

2. Campbell, B. (1966) <u>Human Evolution</u>, Chicago: Aldine Co.

3. Simpson, G. (1942) <u>Tempo and Mode in Evolution</u>, New York: Columbia University Press. Simpson, G. (1967) <u>Meaning of Evolution</u>, New Haven: Yale University Press.

4. Silver, L. (1998) <u>Remaking Eden</u>, New York: Avon.

5. Gould, S.J. (1997). Unusual Unity, <u>Natural History</u>. April 1997, vol. 106, number 3, pp. 20.

6. Tattersall, I. (2000). <u>We are not Alone</u>. Scientific American, Jan., p. 56ff.

7. Gould, S.J. _____.

8. Brown,P.(2004). <u>NATURE</u>, Oct.28, 2004

9. Mayell, H.(2004). <u>National Geographic News</u>, Oct.27, 2004

CHAPTER 8

Population and Habitat

Population, when unchecked, increases in a geometric ratio. Subsistence only increases in an arithmetic ratio.

Robert Malthus
"THE PRINCIPLE OF POPULATION, I."

He kissed me under the Moorish wall and I thought well as well him as another and then I asked him with my eyes to ask again yes and then he asked me would I yes to say yes my mountain flower and first I put my arms around him yes and drew him down to me so he could feel my breast all perfume yes and his heart was going like mad and yes I said yes I will yes.

James Joyce, Molly's soliloquy
ULYSSES

For our game with genes, Robert Malthus is a very important thinker. His first publication in 1778, "An Essay on the Principles of Population, As it Affects the Future Importance of Society, with Remarks on the Speculations of Mr. Godwin, M. Condorcet and other writers", establishes population as a principle. Malthus argued throughout his writings that the happiness and the well being of people were not a consequence of social forms or political movements, but rather stemmed from the primary element of their numbers. He believed that the quantity of people would always outstrip their capacities of production. Population was an initial cause affecting the scarcity of goods, food, and consequently human happiness and social harmony. Population was a principle. Biology preceded culture. Nature held sway over social and political formats, rather than the other way around, as the idealistic and revolutionary thinkers of his day believed.

Robert Malthus was born in the Age of Enlightenment into a family of clergymen and became one himself. The American and French Revolutions took place in his youth. During this period of social revolution and reform, and even later during the Romantic Era, Malthus maintained an Enlightenment approach. He examined human society, its ills, its woes, and its chance for improvement as a set of causes and effects that could be measured and assessed. He developed a position not so much contrary to the idealists, but "gloomier", more "melancholy". [1.]

Malthus' father followed the writings of Rousseau, the Marquis de Condorcet, and Godwin in England. Malthus first essay developed from a debate with his father about those radical and ideal philosophies and formed the basis for his subsequent writings. Malthus reversed their basic thoughts and maintained that human misery and ill did not result from unfair social and political institutions, but rather came from an imbalance between the number of people and the goods available to them. And, this would be so regardless of the political structures. He proposed as a principle that population growth will exceed the means of sustenance. Indeed, his special emphasis was placed on the mathematical distinction that population grew geometrically while production grew only arithmetically. Further, he held that overlooking this law led to speculations, which would create false hopes, idealistic social forms, and even political revolutions. He rejected those radical social solutions as poorly founded, and became a political moralist, as opposed to statistician, theorist, or economist. He proposed that the growth in population be checked by education and prudence in marital affairs, while the increase in production of goods be boosted by constant diligence. An improvement in the human condition would come from these actions, and although social policy might be helpful, from the morality of individuals. [2.]

Malthus notion of the geometric growth of populations has had far-reaching consequences. Indeed, it affects Artificial Evolution, for Darwin read Malthus in 1838 and the tension between geometric growth of population and arithmetic increase in goods helped him develop the biological principles of competition and selection [3.]. Both of these two principles are dramatically at issue in the human population explosion today. Human population, as well as the population of other species, will increase beyond the means of sustenance. And, every population intents to survive and increase. Thus, every population evolves, and at varying speeds, dramatically in some periods or slowly in others.

Since our human population has slowed or halted its rate of organic evolution, while its rate of cultural evolution has sped forward exponentially in a fashion to protect the population from any pressure to evolve genetically, as we detailed in Chapter 3, we stand in a near stable state of expanding population. Our genome is stabilized, but our numbers are growing without much pressure to evolve.

The assessment of future population growth varies, of course, according to the method of calculation. Considerable new work has been done in demography and population studies and much of it revises the doomsday interpretation of sheer numbers. A central theory is that the demographic transition from high fertility and high mortality to low fertility and low mortality that occurred in the post-modern first world is now occurring in the third world [4.]. Thus, global statistical projections must be lower. Other writers hold that the production of goods, and especially food, is already out-stripping population growth and the problem therefore lies in distribution and equity. Others argue that the impacts of populations have been changing. Better ecology behaviors and increasing urbanization worldwide are lessening the harms to ecosystems. The harms of higher human density are smaller than expected. Indeed, there are many advantages to that density, such as increased technological changes, higher production rates, etc. [5. and 6.] Still others hold that social and family planning activities are working and so growth rates will continue to fall. So, statisticians who extend current conditions predict a doubling of today's number to about 13 billion in 2050. Those who believe the boom is over and who site a variety of limiting factors project 7.5 billion in 2050 [7.].

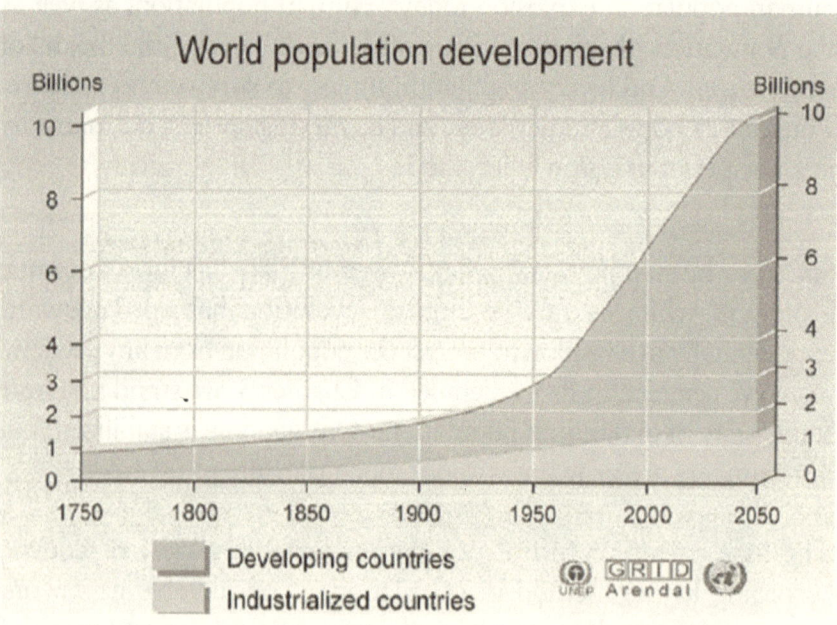

120

The actual number will probably be somewhere in between and the degree of doom will depend upon interpretation. But, the terms of these estimates reveal the central error. The demographic transition, programs for family planning, indeed, the very call for reducing the birth rates, all bespeak a misunderstanding of Malthusian and Darwinian principles. Culture does not dictate biology, rather biological forces precede culture. Here, of course, the forces are functioning in ways different than Malthus expected. The programs and educational efforts are mitigating growth rates, but they are also stifling the mainspring of change. The human population is not growing naturally beyond its means of sustenance. It is not complicating its genome with mutations. Rather, the population and its genome are stabilized. In other words, it is unnatural.

Furthermore, the exponential growth of cultural change is directed toward neutralizing the human environment so as to continually remove pressure for genetic change. All of this is good from an immediate point of view. But, in effect, the primary force of biology is being suspended. Populations should grow geometrically, they should be competing to survive and they should be evolving. Our cultural evolution, in this case, our reduction of population growth, is restricting our organic evolution. It is conditioning change.

Biologists could argue that although this might be true, the forces of mutation and selection would spring forth quickly under any dramatic environmental changes in the future. So, the current status in our genome does not preclude its fundamental biology. Even a population implosion would not affect these underlying mechanisms, because they are laws of nature. On the other hand, the changes in the genome could be consciously chosen and engineered to fit the requirements of a different environment. So, the problem can be approached in yet another and perhaps better way, through Artificial Evolution, for in its implementation the realms of nature and culture cross.

Since the advent of the Industrial Age, the same factors that have caused the population boom have been involved in our cultural evolution. Indeed, the factors are co-extensive, such as increased efficiency in production of goods and food, lower birth mortalities, higher density and closer communication of information, greater nutrition and larger habitat, greater mechanization, etc. Truly, the rates of population growth and cultural evolution have been apace. Both have been exponential.

Simultaneously and consequently, as the numbers of humans increased and the habitat extended under those factors, so the numbers of other species decreased and their habitats have been diminished. While our species has dominated the globe, it has also conditioned the total biosystem to the extent of endangering itself. Some of the most dire doomsday predictions depict the collapse of whole ecosystems and massive extinction of species. In these scenarios, expanding human populations undermine the life systems that they too rely upon, leading to even worse circumstances. Here again, a synthetic line of evolution may be worth examining. Since Homo sapiens comes out of the natural realm like all other species and its conquest of the global environment has been achieved through cultural, rather than genetic evolution, it does make some simple sense to conclude that a continuation of the process is not only possible, but also feasible. To consider the possibility of Artificial Evolution is to admit to the ascendancy of our species, to the domination of the global environment, and to proceed with an anthropocentric, rather than bio-centric, point of view. It is to adopt a technocratic approach to natural situations. And, it does admit the truth: nature is out-of-balance and our nature is out-of-balance.

So, consideration of a synthetic line of evolution is based upon a truthful biological analysis of our situation as the dominant and controlling species. Secondly, it offers an alternative to the current course of matters or, better put, it may offer some assistance to our present course. Thirdly, it takes responsibility for itself. In effect, we got ourselves into this state by technocratic and cultural means, and now we have to get ourselves out of it.

Just how is somewhat sticky. We have mentioned using Artificial Evolution to help preserve endangered species. Today, transgenetic species and cloning are directed toward commercial ends. Various enzymes and drugs are being produced through these new animals because the method is cheaper than other ways. In other cases, attributes are being genetically added to plants and animals to make them more resistant to disease, spoilage, etc. While considerable artificial insemination is being done to aid endangered species, no experimentation is underway to alter those species threatened by the changes to the biosystem in recent centuries. Artificial Evolution could make species more suited to human-altered habitats. Of course, each species and individual is unique, so some might be more adaptable in various ways with characteristics from other species or genera. Methods of increasing numbers of offspring might be devised. Specific changes to vulnerable aspects of certain species could be dramatically effective. For example, frog species could be given tougher skins, to absorb less radiation and chemicals. The social behavior of other species might be altered so that their habitat requirements were different, mainly smaller. In general, the principle is to employ this new mode of evolution to save the threatened biosystem. Now, of course, those species will be different, human-directed and homocentric by definition. On the other hand, they would evolve naturally anyway into something different and, without the option of a synthetic development, they might not survive at all.

If enough effort were devoted to the biosphere basics, perhaps a degree of security could be gained from the possibility of total collapse. This is a proposal for a "New Galapagos" and to envision it we must look centuries forward to see a global environment as a human zone with extensive parklands and protected species. Many of those will be the plant forms required for the basic atmospheric conditions of life. Others will be those of importance for human survival, amusement and quality-of-life. As the dominant species, the one capable of imagining this, even of orchestrating this, we also must admit to being the ones responsible for it. Can we really maintain the current directions of development and still conceive of conserving all the present biosystems and species for centuries indefinitely? It is foolish of us to think so. Therefore, some aspects of Artificial Evolution ought to be seriously considered. A synthetic rather than "all-natural" future is better than none at all.

Just as the opportunities, dilemmas and futures of other species can be developed artificially, so too can that of our own species. *Homo-sapiens novus* can be shaped in many forms. In consideration of population growth, the fertility of another human species could be decreased considerably; or, indeed, engineered in several distinct ways for more or less production. Perhaps our present species could be limited by quartering the ovulation cycles of females or changing the swimming characteristics of male spermatozoa. At the same time, a new species could have other reproductive cycles.

In future centuries, our species will most likely seek to expand its territory. This will take us into the habitation of space and the creation of a new solar culture. It seems reasonable to imagine a new species tailored to these new environmental challenges, rather than Homo sapiens that evolved hundreds of centuries ago to inhabit open grasslands. While there would be some considerable advantage to exporting billions of people to space colonies, it seems conceptually flawed. It would be "unnatural" to maintain the status of our species in a vastly different environment. In such "solar culture" environments, how much population would be desirable? Perhaps the reproductive characteristics of a new species would be variable. Inherent in the idea of an engineered species is that it can be re-engineered and in accordance with the theory . . . very quickly. So, the population growth rates might be variable according specific environmental demands.

In any case, the methodologies presuppose willingness on the part of participants, the vast social contract involved and the bridging of the ethical abyss. The ways and means would unfold toward those ends, if they were undertaken with sufficient wisdom. Nevertheless, the background of this chapter is ominous and urges action. If the statistical formulations are accurate, then we are looking at a global population of over 7.5 to 13 Billion circa 2050, and as many as 26 Billion in 2100. A global society resembling the Hong Kong of today would be utterly untenable. Even increases of 80% every 50 years would yield 180 to 200 Trillion by the year 3000. Such numbers are unthinkable; they promise many population disasters before that next Millennium. An approach of conservation, naturalism, planning and the like is necessary, but may not be sufficient. We must applaud all international efforts at population management. We must exhort all efforts in raising the capacities of production. All the new agricultural methods, all the genetically enhanced crops, all means of efficiencies and greater efforts in equitable distribution must be encouraged. Nevertheless, we must face the fact that population expansion, if not explosion, will continue and that a dislocation, dismemberment, indeed, a disability in the biosphere is very likely. By the end of this century, only 10% of the tropical forests will be left. 50% of the species will be extinct. [8.] Just how this effects the oxygen/carbon dioxide mix in the atmosphere, just how the ozone layer and UV indexes are changed, just how sea water levels and climates are reformed, and just how the surviving species are impacted is all a matter of conjecture now. Certainly, there are many variables and the changes will come slowly in degrees. But, just as certainly, our own species will be threatened. "Homo sapiens is in the throes of causing a major biological crisis, a mass extinction, the sixth such event to have occurred in the past half billion years. And we, Homo sapiens, many also be among the living dead." [9.]

International conflicts to control the most basic resources of water, gas-oil, and food supplies may become overt and hostile. While competition for resources and production has paced population growth and industrialization for two centuries, hostilities may well begin in a conscious way separating the haves and have-nots. Doomsday scenarios are quite possible. Malthus theory now rings true with a death knell. Biology precedes culture. Population does outstrip production, and we are balanced on the verge of population outstripping the sustenance of the biosphere. The factors of population expansion, even when mitigated, of bio-sphere degradation, and of mass species extinction, even given our best efforts, will push the globe into crisis and imbalance by 2200.

An alternate paradigm ought to be considered and put forth. Such a future model would be global, indeed, even solar in concept. It would need to hold basic directives and goals as well as branch options and possibilities. It would need to be a Survival Plan. Broad goals could illuminate the ethical problems and social procedures. As has happened in the past, our fellow species may be the experimental device upon which our own betterment is founded. Thus, the artificial evolution of other species into a new geo-sphere may be the first procedural and ethical step. Our own survival could be based upon those other species. Finally, we could move the whole eco-sphere forward into a new era. While this does seem mighty in concept, the threat of global destruction is real and every positive action we undertake will be worthwhile.

References

1. Malthus, T.R. 1992 <u>An Essay on the Principle of Population</u>, Cambridge Univ. Press, Cambridge, England

2. _____p. VII – XXIII

3. _____p. XI

4. Kirk, Dudley. 1996 Demographic<u> Transition Theory</u>, Population Studies, Vol. 50, No. 3, Nov. , 361-87 pp., London, England

5. Dumont, G. F. 1996 <u>Demography: Unfounded Fears Concerning the Year 2000</u> Politique Internationale, No. 73, Fall , 345-65 pp., Paris, France

6. Peron, Jim. 1996 <u>Exploding Population Myths</u> Heartland Institute, Palatine, IL.

7. <u>CONVERSATIONS</u>. 1997 United Nations Population Division, October

8. Leakey, R. 1996 <u>The Sixth Extinction</u>. Weidenfeld & Nicholson, London.

9. _____p. 245

Malcolm C. McKenna

A lecturer on a tourist cruise captured this view of open water at the North Pole, a sight presumably never before seen by humans.

CHAPTER 9

A New Climate

"Ages-Old Ice Cap at North Pole is now Liquid, Scientists find"

JOHN NOBLE WILFORD
NY TIMES, AUGUST 19, 2000

The expanse of steel-blue water stretched towards the crystalline shapes on the horizon. This open water was not a new sea, but a hole in the North Pole cap nearly a mile across. Observers aboard the Russian ice-breaker, Yamal, including scientists Dr. Claire Parkinson and Dr. James McCarthy, had encountered open breaches in the ice all along their tour of summer 2000, but to find temperate waters at the Pole itself was extraordinary. The immediate impression was a dramatic demonstration of global warming. The story made the front page of the NY Times, 19th of August, 2000. The photo of a pond at the pole said more to the public than decades of scientific analysis about global warming. It appeared that the sky was falling in, and into tepid water!

Articles of alarm, like "Good-by North Pole" [1.], were published and certainly achieved the attention that a catastrophic climatic event should receive. Since 1978, when warming became an accepted issue, the artic cap has receded by about 6%; artic sea water is about 2% warmer; the overall thickness of the cap is roughly 5 feet less. Moreover, the glaciers of Iceland, Greenland, and Europe are melting, and thus, accelerating the heating process by trapping more solar heat in the added water vapors, while reflecting less solar energy from the disappearing whites of the glacial masses. Further, the North Atlantic Drift, that moves warm water north and cold water south may be slowing or reversing. Should that flow stop, "Earth's climate could shift dramatically" [1.].

Broader assessments of the polar water were published during that summer of 2000, such as "Open Water at Pole is Not Surprising" [2.]. The initial alarm was muffled by less antidotal analysis that showed many open areas of long cracks and holes across the ice mass, sometimes including the pole, especially in summer. Indeed, 10% of the zone is liquid in summer and the fissures can be the result of currents and winds. "There is nothing to be necessarily alarmed about. There has been open water at the Pole before. We have no clear evidence that it is related to global climate change" claimed Dr. Mark Serreze of the National Snow and Ice Data Center. Yet, climate models "predict the Artic will be among the first regions on Earth to respond to a Global Warming trend" [2]. Indeed, most of us do believe such a trend is underway.

Our immediate and near future climatic situation may be debatable. But, mainly, this solar event alerted us all to our subjectivity. We are subject to variable conditions beyond our control, even though we assume constancy and predictability. These conditions are global, solar and even universal. And, we are at the will of those macro forces. While warming may be real and drastic in this century, the climate may be cooling or freezing in future or it may be de-oxidizing. There may be many fewer species in this century, but many more in the next century. While some might be lovable, others might be pathogens, new bacterial forms or microbes dangerous to our species. Clearly, the current supremacy of our species is predicated upon many factors, and climate is a major one in determining our adaptive zone. The relationship of species-formation and global cycles does bear thought, because the importance of Artificial Evolution becomes ever greater, the deeper we peer into the future.

Every species evolves, and, indeed, evolves somewhere. Each is adapted to a specific environment and special resource space or adaptive zone in the environment [3.]. When the adaptive zone changes, species are redistributed or replaced by new species formations, and, of course, the main factor in a zone change is a broader change in the environment. While there are other minor factors, the main function of this change is climate. Just as the climate zone supports the life forms, so also do the forms support the climate, such that whole eco-systems exist. Indeed, due to major climate transformations over the ages, there have been many eco-systems. Thus, in the long development of life, evolution and climate have been co-efficient.

One of the fallacies of contemporary thought is that nature is fixed in the nineteenth and twentieth century form. We assume, for example, that Glacier National Park always looked as it does each summer, and thus, it must be preserved as it is. Of course, the nature of the globe is constantly changing. Millions of years ago there were 524 much shorter days in each year cycle, because the spin of the Earth was quicker [4]. Fifty million years ago the Amazon Basin had no rain forest and eight thousand years ago the Sahara was a temperate zone with small lakes [5]. The year will come, long hence, when only 50 days will dawn. The global climate will be very different.

It is curious that the polar ice pack itself has alerted our collective conscious response to climatic change. The experience of the glaciers, of living close to them and dependent upon them, is only 24,000 to 20,000 years back, some 1000 generations back in human history. Indeed, perhaps, we carry some collective response within us. Certainly, these "ice alerts" have struck an inexplicable species response that can not be readily known. Our species struggled to survive through the last glaciations and there may be some "collective memory". Yet, without doubt, the methods of co-operative work, division of labor, future planning of food stocks, agricultural development and livestock management were made in response to the last Ice Age. Climatic change is not only represented by ice, it is actualized by it. Ice has been a dominate motor of change upon the evolution of species. So, it should not be surprising that we react dramatically to its physical presence and its activity. Ice is the most obvious sign for us of climatic change.

Climatic change results from the complex interactions of the Earth's two major heat sources, the Sun and the radioactivity of the Earth's interior. Solar energy is the primary force and it drives both the oceans and the atmosphere. The interior of the Earth drives the tectonics of the continental plates, the position of the land masses, and the flow of the ocean waters and their depths. Climatic variations occur as a result of the interaction of three basic force factors, 1) the amount and distribution of solar energy atop the atmosphere, 2) the composition of the atmosphere, and 3) the character of the surface of the earth itself.

The first factor, the solar energy received, is affected by variations now called the Croll/Milankovitch cycles, after the scientists who identified the phenomena. The Earth's orbit around the Sun varies over a 100,000 year period from nearly a circular path to an elliptical one. This "eccentricity" is a major factor in variations in the seasonal radiation. In addition, the tilt of the Earth varies from 22 degrees to 24.5 degrees in a cycle of some 41,000 years. This "obliquity" results in greater differences in winter and summer seasons and greater total solar energy at the Polar Regions. Lastly, the solar and lunar pull on the equator causes fluctuations in the perihelion, the time at which the Earth is nearest the Sun. This "precession" of the equinoxes has a cycle of approximately 21,000 years. In combination, with variations in eccentricity, changes in precession can yield a difference of some 33 days in the length of summer and winter seasons [4]. (See Charts A, B, and C.)

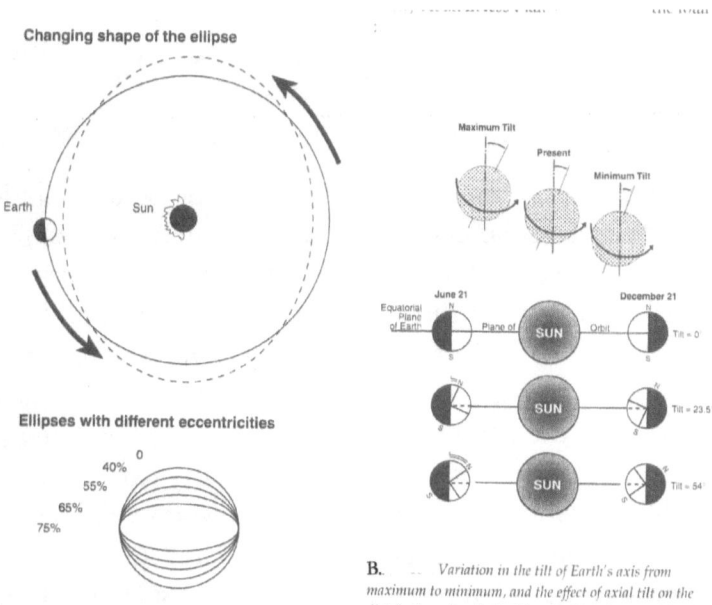

A. ˙. ∴ *The orbit of Earth changes shape from nearly circular to more elliptical. This is termed eccentricity and is expressed as a percentage (after Imbrie and Imbrie, 1979).*

B. ... *Variation in the tilt of Earth's axis from maximum to minimum, and the effect of axial tilt on the distribution of sunlight. When the tilt is decreased from its present value of 23.5°, the polar regions, in summer, receive less sunlight; when the tilt is increased, polar regions receive more sunlight (after Imbrie and Imbrie, 1979).*

133

TODAY

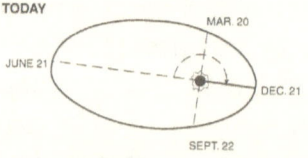

5,500 YEARS AGO

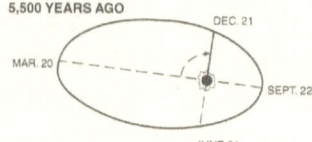

11,000 YEARS AGO

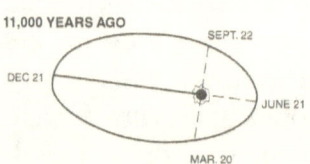

C.   *Precession of the equinoxes. Owing to axial precession and to other astronomical movements, the positions of equinox (March 20 and September 22) and solstice (June 21 and December 21) shift slowly around Earth's elliptical orbit, and complete one full cycle about every 22,000 years. Eleven thousand years ago, the winter solstice occurred near one end of the orbit. Today, the winter solstice occurs near the opposite end of the orbit.*

Precession of the Equinoxes

Solar and lunar torques on the equatorial bulge of Earth cause the time of perihelion (the time at which Earth is closest to the Sun) to vary

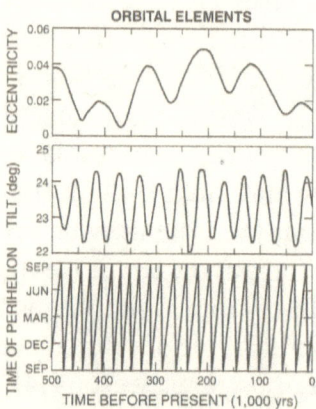

Variations in Earth's orbital elements, eccentricity, tilt (obliquity), and time of perihelion (precession of the equinoxes) computed for the last 500,000 years with a computer program written by Tamara Ledley and Starley Thompson following Berger (1988).

Altogether, the cyclical variations cause the solar isolation to vary. The overall solar energy does not alternate so much, but the fluctuations at high latitudes can be as great as 9% from the mean. (See Chart D.). While these astronomical cycles might seem to indicate a type of large fixed pattern of 100,000 year periods, the Croll/Milkanovitch cycles are complicated by the other two major factors: the gaseous atmosphere and the character of the surface itself in different eons. The very early earth atmosphere was high in carbon dioxide and the prokaryotes were the earliest life forms. When the gaseous elements changed to oxygen and nitrogen, and the eukaryotes came to be, the solar isolation had a much different effect. Also, the amounts of sulfur dioxide and dust resulting from volcanic activity are significant factors.

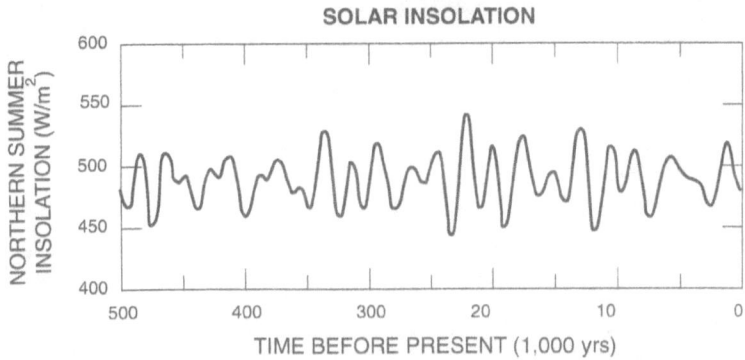

SOLAR INSOLATION

D. Variations in insolation (in watts per square meter) determined from the variations in Earth's orbital elements.

The third major factor is the distribution of the land within the oceans and how the water flows around the globe. The albedo effect is a result of how much radiation is absorbed onto a particular surface in any era of Earth history. And, to complicate matters, the way in which the radiation gets stored is affected by the depth of the oceans as well as polar ice in the particular era. Some ice caps have more or less bubbles, and thus, the nature of the melting differs. Altogether the dynamics of these major factors create a complex pattern of glaciations and climatic change. So, the astronomical cycles are compounded by the inner earth rhythms, and thus no absolute patterns prevail [4.].

However, a general rule of glacial formation and retreat can be seen every 20 to 25 thousand years. The most recent glacial period lasted from 24,000 to 14,000 years ago. Sea levels were 120 feet lower, placing their shorelines near present-day continental shelves [6]. The "greenhouse effect' and carbon monoxide emissions may create climatic dramas in the coming 50 years and may delay the onset of the next glaciations somewhat, but the astronomical cycles and the tectonic/oceanic factors will dominate the climates of the

135

future just as surely as they have in the past. So, the story of our north pole pond is very important in short terms, but astral cycles will prevail in longer terms.

Consequently, the major climatic changes will create new adaptive zones and new species formations. Coming ages will be changed as past ages have been. And, we Humans will be acted upon just as the other species will be.

There have been five major species extinctions in the past related to climatic change. The Cretaceous extinction of 65 million years ago lost 75 % of its species due to the asteroid impact. The Permian extinction of 245 million years ago lost all but 4 % of its aquatic species. Other extinctions took place 208, 367, and 439 million years ago [4.]. Many biologists and social scientists calculate that the sixth massive extinction is underway [7.]. This one is caused by enormous habitat destruction, particularly in the tropics, over-hunting, and chemical pollutants resulting from our population growth and our use of the global ecosystem. The rate of species loss is comparable to other mass extinctions; however, the rate is located in specific limited ranges of the rain forests [7.]. While this may be lamentable, the long term impact on Humanity may not be as great as its perception is today. Or, it may be as disastrous as the worst forecast. There simply is no solid predictive model.

The central question is not that the nature of nature is changing, but rather "how do we wish it to change?", or "What place do we want to call home?" The immediate and dire problems of the next 50 years will be solved through conscientious common global actions that involve our collective will. Indeed, such solutions already involve Artificial Evolution, but an overview of global climatic forces does direct thought to our distant future, and our new evolution can be an important tool in creating our future home.

Of course, the next generations will not be greatly impacted by Artificial Evolution; there will probably be little impact in the next 10 years. Yet, the 30th generation from now will certainly be affected, by the year 2900, and, the situation for those future generations looms large. All organic evolution will be interdependent with the global cycles we have identified.

If we humans can admit to our role as prime actors in nature, if we can assume a broad anthropocentric view of the future and find confidence in our abilities to collectively solve the immediate crises, a new era begins to appear. It is marked by operations that the words collective, interrelated, networked, electronic, and super-organismic, only to begin to name. Gregory Stock proposes METAMAN [5] to title the interphase of "man with machine". There are many pop-culture icons that both mimic and perhaps predict such new forms, like "Ironman", "super-heroes", etc. This new era then will have a new entity, a new global life form, or a multiplicity of them. They will live in a future called Space Culture.

Contemplation of global climatic cycles leads us to realize that the character of this new era is not a romantic exploration, a commercial apology, or merely Sci-Fi. It is rather a survival requirement for our future. It is senseless to argue for the preservation of current states in nature. Change is underway and inevitable. Of course, we must, and indeed are conserving wild populations, vulnerable species and habitats. But, the overall view needs to be directed to "What kind of globe shall we call home?"

Will the Earth in the immediate coming Space Era be one of clear waters and fresh air, one of gardens and diverse wildlife, one of integrated populations? Yes, indeed, because such developments are already underway, but, more importantly, because the processes of conservation and adaptation are not mutually exclusive, or at odds.

From the view of global cycles we can see our species and our situation in a broad scope. Clearly, the processes of stasis and conservation are antithetical to the solar and universal forces creating change across the Earth. The processes of adaptation and mobility are aligned and harmonious to those same universal dynamics. The way of nature is not calm, fixed, and contained; rather, it is chaotic, conflicting, tumultuous and mobile [8.].

We should recognize these principles of adaptation vs. preservation, or mobility vs. stasis. These fundamentals, adaptation and mobility, will bring us across the generations ahead and beyond the immediate crises. The significant difference is that collectively we possess a new power. We can rearrange the life forms to design them for specific, changed habitats. Artificial Evolution is central to the climatic changes, astral dynamics, and environmental shifts here on Earth in our very own adaptive zone.

In our new mode of evolution, natural selection is no longer the only directional force in nature. We can reform our own and other species. If arguments from the character of our cultural evolution, from an anthropocentric future, from the population explosion, from the Utilitarian Principle, or from the future generations debate were not enough, then an examination of our global adaptive zone in climatic cycles alone would favor the advantages of Artificial Evolution. It will be required for our survival. Most probably, it will be a central methodology for our species in the Space Era to come.

While this may seem far-fetched, exotic, or futuristic, there are many examples of our collective actions that have dramatically affected our global situation. The successful completion of the Panama Canal involved the efforts of the peoples of Central America, Africa, China, France and America. Over decades and many obstacles, a passage that was closed eons ago by the collision of continents was opened by humans working together. It established the supremacy of American engineering, and in many ways contributed to the industrialization that has caused a speed up in our cultural evolution. Indeed, the very global warming we have discussed has been enhanced by the passageway. Another example would be the process of reducing infectious diseases. From Pasteur and Sauk to the present day scientists, many nations have worked on that goal and continue to work on reducing inherited disorders. As we have noted, problems with our population boom have come along with the benefits of reducing diseases. Yet, certainly, both of these collective endeavors are much greater than the engineering required to manipulate species into new climate conditions. Actually, many of the processes are in practice today. If climatic forces demanded action, and if it became a collective imperative, we could adjust numerous species to new zones. Indeed, we could adjust ourselves, and we might be the critical species to protect the others.

Thus, the ice that has so many variations of blue, of bubbles and buoyancy, of transparency and luminosity, the ice that has such a hold on our attention and old memory, is much more than a fascination, or symbol of Earth. The ice is, and will be, a barometer of climatic change for us. Just as the ice is now melting, and holds our focus upon its physical condition, so too will its reformation alert us to the next glaciations, which will ensue some 1,000 years from now. Long before that time, we as a species will have mastered the new techniques of Artificial Evolution and employed them for ourselves and other species.

A future culture and a future landscape can be cultivated and planned, just as we have done crudely and slowly by artificial selection in the past. By adopting the principals of adaptation and mobility, rather than holding to conservation and stasis, we will finesse the climatic and habitat forces that have condemned so many species in the past. Truly, without these principals and without Artificial Evolution, our species might not be able to survive the centuries ahead. However, acting together as a wise species, we can create the garden park that our globe can be.

References

1. Eds. "Goodby North Pole". Earth Island Journal. Winter 2000, Vol. 15, p.4.

2. Wilford, J.N. <u>Open water at Pole is not surprising.</u> NY Times, 29 Aug. 2000.

3. Mayr, Ernst, 1988. <u>Towards a New Philosophy of Biology</u>, Harvard Universtiy Press, Cambridge, MA, p. 135FF.

4. Barton, Eric J. 1996. <u>Climactic Variation in Earth History,</u> University Science Books, CA.

5. Stock, G., 1993. <u>Metaman</u>, Simon and Schuster, NY, p. 175.

6. Crowley, T. and North, G. 1991. <u>Paleclimatology</u>, Oxford University Press, England, pp. 256ff.

7. Leakey,R. 1996. <u>The SixthExtinction</u>. Weidenfeld & Nicholson, London.

8. _____. "New Eye On Nature: The Real Constant is Turmoil." NY Times, 31 July, 1990.

CHAPTER 10

Bio-Utilitarianism

The worth of a State, in the long run, is the worth of the individuals composing it.

John Stuart Mill,
<u>LIBERTY</u>

Ethical deadlocks often stem from the assumption of differing ethical systems. Religious orders have various ethical positions, while the secular traditions all treat issues differently too. Buddhist, Christian, Libertarian, Contractarian and Utilitarian ethics all have contrasting approaches to problems. In America, as in England, the common sense view of right action descends from liberal English philosophy of the last century. This common sense approach can enlighten our views of Artificial Evolution [3] and especially its function of germ-line engineering.

The main proponent of liberalism was John Stuart Mill (1806-1873). *Utilitarianism* (1863) defined the ethical philosophy that we now know by the name [6, p.893]. Mill wrote that "utility, or the happiness principle, holds that actions are right in proportion as they tend to promote happiness, wrong as they tend to produce the reverse of happiness. By happiness is intended pleasure, by unhappiness, pain." [6, p. 900]. He then defends this principle against Epicurean interpretations, against accusations of pleasure in general. Mill writes further that "the utilitarian standard of what is right in conduct, is not the agent's own happiness, but that of all concerned". "In the golden rule of Jesus of Nazareth, we read the complete spirit of the ethics of utility." And "to promote the general good may be in every individual one of the habitual motives of action." [6, p. 908]. In our modern democracies, this crystallization of the common good still conditions how we think of right and wrong, of what is just and fair. The golden rule continues to inspire action.

More specifically, the notion of the common good has grown to include our common heritage. Most Americans and Europeans consider our historical tracts and monuments to be held in common. So, too, our national parks are a common heritage. Indeed, most Americans conceive of "nature" as a common property and its experience as a common right. Even health and safety are common rights. It is upon this basis that the environmental movement has laid claim to the global ecosystem as a human heritage. The well being of the whole and the common good of all people must exceed the rights of any individuals. The globe has become our common home and its preservation our common imperative.

A very similar chain of claims is made about the human genome. It is considered a common heritage because the well being of our species resides in the genome, and, therefore, its preservation, like nature, is an imperative. Both the ecosystem and the human genome are viewed as common property and tantamount to the common good. The principle of utility is then applied to both of these natural biological entities: the good of all lies in their preservation.

However, the processes and actualities of biology are quite different from those of sociology, politics, economics, or mathematics. "Nature" and "the genome" are useful as abstractions for debates in various disciplines, but neither has any biological foundation. There is no "nature" or "genome" in biology. There are only individuals that can be thought of together in biology, only individuals that can be considered collectively as species of individuals [5, p. 224-346].

Consequently, the application of Utilitarian principles to biological entities needs review. How can we view the common good and the happiness principle in the social context of the collective global ecosystem, the collective genome? Indeed, what are these social constructs and how are the claims to them made?

The biological characteristic central to both is change. The abstract, "nature", is compounded from individuals of manifold species in dynamic balance in complex ecosystems. Each individual is striving to survive and reproduce and is selected to do so. Change in the environment brings a concomitant change in the living individuals. The individual genotypes change. The principles are adaptation and mobility, not conservation and stasis.

Nature and the human genome are abstractions constructed for social debates. Since change underlies the biology of both social constructs, it follows that the common good and the experience of our collective pleasure will be best served when aligned to the forces of adaptation and mobility.

Obviously, this is not to argue for the demise of baby seals or the wasteful destruction of habitats, etc. Rather, it is to claim that the common good can not be served by fixating or making static those biological processes that are dynamic and mobile. Conserving nature is not in the long term good of the ecosystem or humanity. The Utility principle in biology will be enhanced when the greatest long-term human good is aligned with adaptation and change. From a biological perspective, every species is best served, is happiest, when best suited to it s adaptive zone. Actually, this is what it means to flourish, and why there is evolution – the common good is the best adaptation to a specific environment zone. Each member of every species is happiest when best suited. And, in biological terms, if not sociological, political, legal, etc., Home Sapiens is and will be happiest when best suited. Our greatest common good then appears as our suitability to a zone. The conservation of our common genome seems then as incorrect as the long term conservation of the global ecosystems. In truth, both are in dynamic change.

When we examine biotechnologies with this point of view, many moral imperatives are reserved. The common sense tradition of utilitarianism argues in favor of procedures otherwise "weird "or "unnatural". New bio-actions are right and justified to the degree that they will result in the greatest happiness, defined biologically as the greatest suitability to future environments. Such environments will change and to achieve our greatest good we will have to change as well. We will have to evolve.

In-vitro fertilization processes, dubbed "test tube babies" by critics, would gain approval when judged by these ethics, because the procedures are equally as safe as normal birthing methods and a whole set of parents are helped. These people get to have families and the overall good is served well by their happiness. The parents are individually aided and the common good benefits from their happiness. Of course, from the point of view of another ethical system with a different set of assumptions, this procedure violates the sanctity of life. The parents are happy, the in vitro babies are happy. Society is happier and more functional with happy families, so it passes on Utilitarianism principles. And, it shows how capable utilitarian principles are in supervising the techniques of the gene game.

Now, cloning persons presents complex ethical questions. As we noted, clones are almost like twins, separated by considerable time. Large scale sets of clones are so different in degree that they become different in character. The Bragus bull clone of 1988 was like nothing in nature. "Dolly" also was like nothing in nature, because she was twinned by her adult self, which is impossible and only conceivable by techno-minded humans. As we asked earlier, "Was Dolly true to her kind"?

Yes, indeed, if based on the utility principle. Sheep are domesticated in our service and "Dolly" was born to help continue herd characteristics and maintain the genetic information she holds. The characteristics help us in various ways so the benefits promote the common good.

Cloning humans is less clear. Obviously, we are not domesticated to the service and the utility of other people. Sets of clones violate the whole notion of human dignity. But, individual situations are more compelling.

Cloning the stem cells of an individual might be possible in the future. This would allow organs compatible to the person to be developed as transplants. Truly wonderful medical benefits might come from this and such a process would benefit the common good. Further, the DNA would be preserved to make a twin if the original child were lost, murdered, or killed. This might benefit families and contribute to the good.

Gay families might also benefit from this process. Each member might twin themselves to create a family and, while it is somewhat odd to imagine, the larger social fabric is more tightly woven when family structures are happier and more functional. Again, utilitarian ethics are a sound guide.

These are really special cases of human cloning. So, what of expanded programs? What of cloning for other reasons? Or, for future environments?

It may come to pass that people will be in service to space programs or social programs that require extended time periods, even much longer than a lifetime. There may be space locations that require staff over centuries of time. We may have been able to extend life spans by then, but, on the other hand, we may be able to have a "twin" take over a role just as well. The general good will be served by the person maintaining the same genetic make-up to do the remote location job, just into another generation. This would be like "Dolly", continuing the genetic information generation after generation in service to the greater good of the mission. We readily accept the sublimation of the individuals to the general good in various situations, and, so, this does not seem to be an unlikely stretch. It would be a program decision for a broad purpose. It shows another implementation of human cloning that may well become acceptable, supported by utilitarian ethics.

Reforming human characteristics is an even broader ethical situation that has been addressed at length [7] [10]. This procedure of transgenetic operations on humans is the same as that currently and broadly in practice on many other species of animals and plants. Variously called engineering the gene line or redesigning the genome, all the techniques are instances of

Artificial Evolution, as are those common to animals and plants now. This brings up an underlying ethical question: Why is our species different?

We should remember that assumptions about "Nature/Human" underlie the debate. The whole argument swings around "are we part of nature or outside of it?"
The ethical standards of Utilitarianism ask directly about the good common to all humans. As we have just argued, these standards strongly support the notions of adaptability and survivability into new zones.

There are specific possibilities that can illustrate the issue and prove the principle. The particular points concerning each of these operations can be tossed back and forth at length. We must remember that, although the specific techniques of Artificial Evolution do, of course, require thought and review, each one is an instance or example of our broader mode. Some techniques may be too costly, others too unsafe, others plainly wrong, and still others quickly passing to other methods. Therefore, our ethical attention must be focused upon the whole process of our new evolution and the purpose of the gene game.

We should recall the new Laws to better attend to this:

The Law of IMMEDIACY	Not Darwin's GRADUALISM
The Law of TRANSGENIC DESCENT	Not Darwin's COMMON DESCENT
The Law of SYNTHETIC SELECTION	Not Darwin's NATURAL SELECTION
The Law of ARTIFICIAL EVOLUTION	Not Darwin's EVOLUTION
The Law of MULTIPLICITY	As Darwin's MULTIPLICITY

These Laws and their process will devise not only whole new types of beings but a whole new way of being. The character of our species is changing and furthermore, its purpose is changing. We humans now have a new role in our corner of the Universe. Indeed, as we shall see, the modifications to the Laws of Evolution open a wider role for scientific knowledge as well as a wider role for our species. We are bringing purpose, human progress and history into nature for the first time.

Surviving is the most basic urge for every species. This factor is primary for the happiness of all humans. Second comes suitability. We will all be happier as a whole when we are most closely adapted to our specific zones, which we have identified as Cultural/Natural Environments. The common good is best served when these factors are met. Of course, lodging our species in an immoral future would not contribute to mutual happiness. We are also suited socially and psychologically, so these unusual future horizons must include ethical agreeability and appropriateness. Obviously, the growth and development of our species into multiple forms can only be managed by our mutual agreements and democratic reviews. Thus, the social concord would be a third factor conditioning the common good for our growth into future zones. Finally, a fourth factor would be the social contracts involved, or the exact process of implementing these changes. As we have discussed, we can fairly put aside the massive "program" idea, because the process will be governed by incremental individual decisions at first on the family level and then on the institutional level. Elective surgery will be the principle, rather than state mandates. Slowly, new groups of people, new sub-species and species, will be established in accordance to our survivability, our suitability, our social concord and understandings. This is the game of the genes. Our species is expanding and evolving just as Hominids past have done, just as all species do. We are in charge of the process with new capacities, responsibilities, and, most of all, a new vision of our purpose.

When we imagine the environments of the future we extrapolate climactic factors, such as the Croll/Malankovich cycles [1] and some immediate ones like ozone depletion and habitat destruction. However, it is our cultural environment that is so much more dynamic and immediate. While our organic evolution has changed little in the past 40,000 years, obviously our cultural evolution has been extraordinary. Indeed, while genetic evolution changes steadily, but without clear patterns, our cultural evolution grows exponentially [8 & 9]. The accumulated technological developments of the last two centuries, then, only indicate how vast our cultural environments will change in the future centuries. We are increasingly living within this created cultural environment [4], and as a consequence, our evolution will need to be paced to these cultural environments. The greatest common good will be conditioned to these environments as well.

It is important to note that we will devise numerous cultural environments at the same time. Clearly, the expansion into space will offer possibilities for specific adaptive zones, as each space location will have a specific purpose and condition. Freeman Dyson holds that "In the next hundred years...we will see genetically engineered plants and animals adapted to the colonization of various asteroids and planets....Long before a thousand years have passed, life will have spread over the solar system...As humanity expands its living space away from the earth...our one species will become many...some adapted to heat, others to cold...some to high pressure, others to living in the vacuum of space." Dyson envisions the multiplicity of species continuing in our hands, as Artificial Evolution. "Compared to the slow pace of natural evolution, our technological evolution is like an explosion." We will create a world "that spins a thousand times faster." [2, p.147-173]

From this point of view it becomes critical not to question and debate Artificial Evolution, but to press forward its particular operations in order to keep pace with the technological developments what will bring us across the solar system. While there may be some utility in devising the globe as a repository of natural life forms, our greatest good will be served by adapting to the many environments of the future. Our harmony and greatest happiness will lie in our suitability to those environments.

It appears then that the liberal utilitarian ethics described by Mill, which underlies our common sense view of right and wrong, support Artificial Evolution and its main operation of germ-line engineering. Indeed, if the greatest utility is to be served, this ethics encourages the process that will keep our species apace with our cultural development and in harmony with the nature of our expanding environments in the future.

We can call this ethics Bio-Utilitarianism. It will serve us as the way to judge both the individual techniques and the whole process of new evolution. Our game of genes will be served by this systematic ethics. As we have seen, it is at the basis of our liberal democracies and it guides our common sense notions of particular rights and wrongs. Clearly, this broad ethical method can be the control factor for our evolution into specific environments of the future.

Bio-Utilitarianism can help us determine our step by step development as we align our species with the universal forces of change, adaption and mobility. As we noted, the development of our species will be paced by common family values and historical institutional practices. Bio-Utilitarian ethics resides comfortably within that same history and will serve well into the future. To survive and to flourish, our species must adapt into our own cultural environment and into the Space Era before us. This ethical code can be our guide.

References

1. Barron, E. 1996. <u>Climactic Variation in Earth History</u> <u>University Science Books</u>, CA.

2. Dyson, F. 1997. <u>Imagined Worlds</u>. Harvard University Press, MA.

3. Ennenga, G.R. vB. 1997. "Artificial Evolution" in <u>Artificial Life</u>, M.I.T. Press, Vol. 3, No. 1.

4. Hardison, O.B. 1989. <u>Disappearing Through the Skylight</u>, Penguin Books, N.Y.

5. Mayr, E. 1988. <u>Towards A New Philosophy of Biology</u>, Harvard University Press.

6. Mill, J.S. 1863. "Utilitarianism", <u>English Philosophers From Bacon to Mill</u>, 1938, Burtt, E. Ed. Random House, N.Y.

7. Silver, L. 1997. <u>Remaking Eden</u>, Avon Books.

8. Simpson, G. G. 1942. <u>Tempo and Mode in Evolution</u>, Columbia University Press, N.Y.

9. _____, 1967. Meaning of Evolution, Yale University Press, CT.

10. Stock, G. 2002. <u>Redesigning Humans.</u> Houghton Mifflin.

CHAPTER 11

A Future Landscape

Three quarks for Muster Mark!

James Joyce,
FINNEGAN'S WAKE

{*This is the origin of the name for the physical particle, Quark, proposed by Gell-Mark and Zweig.*}

The spirit of wonder which led Blake to Christian mysticism, Keats to Archadian myths and Yeats to Fenians and Fairies, is the very same spirit that moves great scientists; a spirit which, if fed back to poets in scientific guise, might inspire still greater poetry.

Richard Dawkins,
UNWEAVING THE RAINBOW,1998 .

Artificial Evolution has given us many transgenic plants and animals. Recent advances in cloning from adult mammals will yield wider distributions in these larger animals. All of these beings have been developed for commercial reasons, for drug production, for longer shelf lives in markets, for more stabile herd characteristics, etc. Indeed, as we have revealed, this expediency has made Artificial Evolution and opened our way beyond natural evolution.

As we contemplate the ecosystems of the future here on earth, we must forego the notion of fixing the 19th century view of "nature" in national and global parklands. Ecosystems must evolve. We really must work toward achievable goals. The 24th century will indeed have beings other than our species, but they will be linked immediately to our habitats and adapted to them. Nature preserves will be small islands surrounded by our much expanded space culture. They will be akin to museums or terrariums. Preserved species will exist in what can only be termed "dioramas". They will serve as base-stock and amusements from the natural past. The vital forces of living will have passed into Artificial Evolution.

Yet, this new mode of evolution can be used to aid the endangered species of today. The environmental movement has not taken up this challenge as yet; indeed, few have seen the possibilities.

In 1998, there were 668 species of plants and 458 species of animals on the endangered and threatened lists. Of course, there are hundreds more proposed and added since, and probably just as many that have disappeared before we even recognized them. Details of the listings are updated at the International Union for the Protection of Nature (IUCN) web sites:
http//w3.iprolink.ch/iucn/info_and_news/index.html.

Perhaps more importantly, Conservation International has designated 19 bio-diversity regions as hot-spots for species and habitat destruction. As these hot-spots are harmed, so the endangered lists and extinctions well grow. The list reads as the most fascinating areas of our Earth, where the plants and animals are not only the most verdant and magnificent, but also the most interconnected.

Of course, every effort must be made to retrieve this process of destruction. We must halt the incursions of our species into these threatened zones, prevent the growing expansion of our population, and protect these bio-zones. However, we must also recognize that this is all rear-guard action. All our very best intentions and efforts will still bring us to a 24th Century very unlike the 19th. The Central factor in saving this bio-diversity is a recognition, or an admission, on our part. We must admit that we are the central force in this destructive process and then move on to recognize and admit that we will be the central force on our globe and in this solar system. We must not only take the blame, we must also take the responsibility. Earth is sapiens-centric. And, the Solar System soon will be.

While there are manifold extinction pressures, destruction of habitat is a main cause of the current species reductions. Numerous species might be saved, granted in a modified form, if we could understand how to redesign their territorial requirements and/or their use of their zones. Enhancing their reproductive capacities might help in some cases; changing feeding habits might be effective in others. Extensive research, as well as trial and error, would be involved. Of course, critical decisions about which species to choose would loom large. From this conjectural standpoint, it is clear that Artificial evolution could assist in the survival of many species. This is a very new idea, so it must be treated as one. It must be taken up, developed and nurtured by specialists. Even though our hypothesis must be a general one, the actual process would be an exact one, made-to-measure, and designed for specific ends.

Yet, here we can merely speculate, hypothesize, and, indeed, fantasize. We have five examples of what might be possible in our future landscape, our future solarscape. Here, you readers must imagine scenes in the 24th Century, where city states are built below grade, where numerous space habitats extend our range into the solar system, where planet Earth is sapiens centric and a diorama, a parkland. This is not meant to be specific or concrete, but to be suggestive and inspirational. All the examples are hypothetical, and, truly, fantastic.

Some are derived from the illustrations in the Complete Encyclopedia of Illustrated Animals, by J. Heck and Others current it Darwin's time. They often accompany the reissue of Charles Darwin's *Origin of the Species* that was written so short a time ago, but, culturally, so long distant. The Condor and the Tortoise were illustrations to the original first edition, 1859.

Transgenetic Vulture-Condor

The California condor is very endangered. Our vulture can stand in for the condor. The endangered species is genetically trans-classed by clipping in the capacity to digest vegetable and fruit matter, rather than carrion, and a larger egg production to increase numbers. So, the condor of our future soars above the agricultural field and groves of California, cleaning up on vegetal matter, somewhat modified, but flying again.

Our second example is good stuff for future space locations. The plague of the past becomes a bountiful food source in the future. Our transgenic locust is enlarged two-fold. The wings are shortened and the hind joints are reduced to prevent mobility. The body and upper legs are enlarged to yield more protein for consumption. The genes that protect crickets from freezing in glaciers and cold zones are clipped into our locust to let it survive through freezes. This is a grazing beast to eat up the molds and low form vegetation that survives in solar telecenters. It is a high protein food source for Homo sapiens novus in that locale.

Plate XXX. Locust

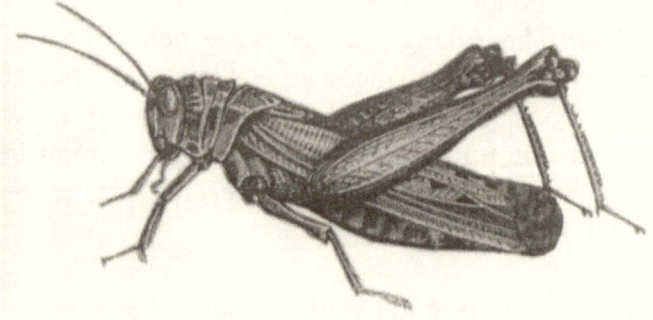

Transgenetic Locust

Number three is a similar adaptation to provide food. The tortoise that provided meat for sailors exploring the globe does the same service for explorers in the future. Our tortoise has longer legs for more agility and an expanded body to develop more meat. Here is another vegetation beast able to survive various conditions and provide vital support to the humans of space culture. Low gravity locations would be ideal for our modified tortoise, because it would be more mobile.

Plate XIII. Tortoise

Modified Tortoise

Our fourth example is both a transgenic survivor into the future and an invaluable cyborg worker. The African elephant must have smaller range in future, a greater use of its zone, and be less destructive.

Here we suggest a smaller body form, without tusks that can be so destructive. In addition, bovine genes have been added along with an improved digestive tract. The elephant becomes more of a plains grazer.

The extraordinary natural brain power is given computer assistance and radio receivers linked directly to the cortex. The cyborg elephant can be given commands in remote locations from satellite stations. Other genetic modifications might be made to their hide, to their abilities to survive hot/cold zones. Clearly, this transformed creature could enjoy low gravity environments, where it would be almost agile.

Such a transgenic future could both save the elephant species and provide as valuable assistant to our own future species in specialized locales.

Plate IV. Elephant

Transgenetic Cyborg Elephant

The last example is from our natural cousin, the chimpanzee. Nimble, intelligent, and naturally formed to be the ideal size and weight for space mission, the chimpanzee may well become our greatest ally in the exploration of our space niches.

The species can receive many of the modification of other mammals, but it is obviously well suited to have AI links. Our example shows the audio AI and visual AI ports added to the natural channels of site and sound. Thus, the species is equipped, like our cyborg elephant, to receive information from satellite links, telecenter to telecenter (Chapter 10). Just as our own species, the chimpanzee could enjoy an immediacy and sense of community reaching across our solar system centuries hence.

Transgenetic Cyborg Chimp

These examples indicate the direction our new mode of evolution might take us. It is indeed fantastic. And, it is rude. It is an affront to the dignity of these creatures. It is even an affront to our humanity to imagine redesigning ourselves for the habitats of the future. But, the argument here is more basic, more dramatic. This is survival. We will change. All of these species will change or expire. We must act to align ourselves with universal mobility and change to thrive into new zones. The species that survive will grow and change with us. Like the planet, like the solar zone itself, they will be sapiens-centric.

Chapter 12

Space Culture

"All is full of what is."

PARMENIDES, 5TH CENT BC.

SIMPILICUS, *PHYSICS*

Space culture will be what we make it. And, more than in any culture of the past, the decisions to shape it will have to be carefully thought out and designed. Moreover, there will be many cultures tailored to specific locations in space and beings will need to be made to fit into those specific cultures. Space culture may be just beyond our current horizon, but its features can be seen clearly now.

It will have many of the characteristics of our early 21st Century culture, but accelerated. Marshall McLuhan identified significant phenomena of the late 20th Century that certainly will be carried forward. In the "Age of Implosion" [5,]as opposed to an age of expansion , he writes: "Electricity has wrapped the planet in a single cohesive field or membrane that is organic, rather than mechanical in nature...Habits and attitudes natural to centuries of expansion now yield with equal naturalness to the intense pressures of an electronically unified world." He makes the critical distinction that the Age of Discovery was succeeded by the industrialized, mechanized revolutions of the 17th and 18th centuries, but the Electronic Age of the 20th has collapsed the globe by media that are organic and condensing, such as, television, movies, international news, pop culture, music, especially mass distribution and electronic advertising. All these processes unify, standardize, and quash distances at electric speed. Space culture will be expansive and exploratory, like the Age of Discovery, but propagated by the same agents of implosion that shrunk the globe, unified it, and made of it a membrane, a web.

The process, then, will continue on another level, indeed, into another zone. Personal computers, digital processes, the internet, and satellite links enable individuals to participate in broader networks that bring us world culture at greater speeds with a lesser need for direct experience. In space culture, the organic membrane of our media will grow and slither about in space. Immediacy, collectiveness, transitivity and virtuality will be its nature and its method.

McLuhan continues: "when information movement speeds up a great deal, center-margin patterns yield to centers-without-margins" [5]. In industrial culture, margins were isolated areas of specialism and fixity, for example, the outlying village or colonial towns where goods mail and fashion arrived last, sometimes years later. The colonial societies remaining today, such as in the Caribbean are now finally being integrated into the membrane, the web of implosion. Innovation and communication will be centered everywhere at once in space. Simulcast will be normative. "Decentralism today is the child of space-time and instantaneous information movement". [5].

The means of information movement, its simulation of other experience , and its excited, bombarding enforcement by electronic sight and sound have made full real time and real space experience unnecessary, even if desired. Our senses of taste, touch, and smell have been percussed by electronic imaging. Indeed, even seeing and hearing are no longer required for us to be experiencing and accepting as real that which is electronically transported. The virtual stands in for the real. Millions have traversed Mars by telecom aboard the Martian Rover. Collectively and virtually, humankind has crawled upon Mars. This very same form of experiencing will be daily and basic to space culture. Our common practices, all that we wish to share, will be everywhere virtual. So, oddly enough, the expansion of our species into the blue - black surround of our solar system will be marked by electronic immediacy and a foreclosed reality, shared as virtual, as a common sheath.

Paul Virilio has written of these forces extensively. "With the interfacing of computer terminals and video monitors, distinctions of here and there no longer mean anything." [7,] "Chronological and historical time, time that passes, is replaced by a time that exposes itself instantaneously. On the computer screen, a time period becomes the support-surface of inscription. Literally, or better cinematically, time surfaces" [7, p.14]. "Speed distance obliterates the notion of physical dimension. Speed suddenly becomes a primal dimension that defies all temporal and physical measurements." [7.]

This is the primal dimension that will cauterize space in our future society. There arises "a new type of concentration: the concentration of a domiciliation without domiciles, in which property boundaries, walls and fences no longer signify the permanent physical object." [7.] The nodules in the membrane of space culture, its centers-without-margins, even its populations, will be imploded by its means, its media of transport, of communication, and of representation. Space culture will be imploded by its way of experiencing.

"Depth no longer involves the visual horizon, nor the vanishing point of perspective. Depth now pertains exclusively to the primitive grandeur of speed." "The center of the universe is no longer the geo-centric earth or the anthropocentric human. It is now the luminous point of a helio-centrism, that special relativity helped install" [7,] With this in mind it becomes ludicrous to imagine the culture of the coming millennium as positions in the solar system, as stations or orbs about the sun as in a child's science project. The stations, locales, habitats, walls and architecture in mass will be imploded, their peoples enmeshed like fish schooled together in the sea.

"In this era of instantaneous, interactive imaging and an interfaced architecture based upon luminous centrism, we find an obligatory interactive confinement, a kind of inertia of human population, for which the term *teleconcentrism* might be proposed." Virilio continues, "...the geo-morphological unity of the state dissolves" [7]A new form of future unity appears at the turn of the year 2,000. The instantaneous spacelessness of implosion will pace our expansion into space. The organic electronic membrane will be humane. We will become akin to it, immersed and woven into its tissue. The space, the matter, the locales, and the gravity all melt into media of movement, virtuality everywhere at once. The primal force of speed will underlie all experience. Unity will extend across space culture in nodes and tendrils that have been interwoven without margins. The facades and structures of this society will be not so much transparent and reflective like the sky-scrapers of the 20th Century, but luminous. Farther off will be no farther away and, curiously space culture will have no space.

All of this differs radically from the concepts about space in the 20th century. "Remote habitats" have been conceived. "Colony" is the most prevalent word to describe the effort to expand, as in "space colony"; the distant "colony" or the "outpost" has been depicted because the time periods of thruster travel establish a physical distance, the very distance, as we have seen, to be squelched by the media of space culture. The remote "colonies" have been imagined as though they were distant foreign shores in the past Age of Discovery. They are seen as isolated from the mainland, from the center, from the home base of Earth. Book titles, like "Islands in the Sky" [6.], display this misconception. The ideational ground work for establishing "outposts" on the Moon was contradicted on the first mission, as the lunar landing was televised back to us. As the long distance feat was celebrated, the astronauts spoke across the ether to Earth immediately. The connection was instantaneous, revealing the critical aspects of our new culture in our first step away from the planet. The truth of this relationship will be repeated into our long future. "Colonies" and "outposts" will be replaced by something akin to muscle fibers in a continuous membrane. The centers will be nodes or synapses, sinuous rather than disjointed and remote. Space culture will flex electronic, muscled by a lithe sheath. The third dimension will be in demise, regardless of the thruster travel times involved, because the space will be compressed. All will be virtually imploded and the language will be Luminosity. Space culture will teem with tele.

Thus, a central notion of the last century can be discarded. Locations in space will not be "remote colonies" administered from a center. Indeed, a new description is needed: space cells, or nodes, or telecenters can be proposed.

Yet another more significant projection and misconception lies hidden in notions from the last century. The language calls for "manned missions", usually following upon mechanical unmanned lead missions. The robotic silicon scouts lay the land out for man to explore. If our analysis and forward vision is solid, then surely, major questions develop, such as, what kind of beings will live in space culture? Better still, what beings can live there, in the space of no space, within the membrane of our future?

Photo by Leonard Rue Enterprises

A SUPERBLY WELL-ADAPTED CRITTER,
THE AMERICAN PRAIRIE DOG.

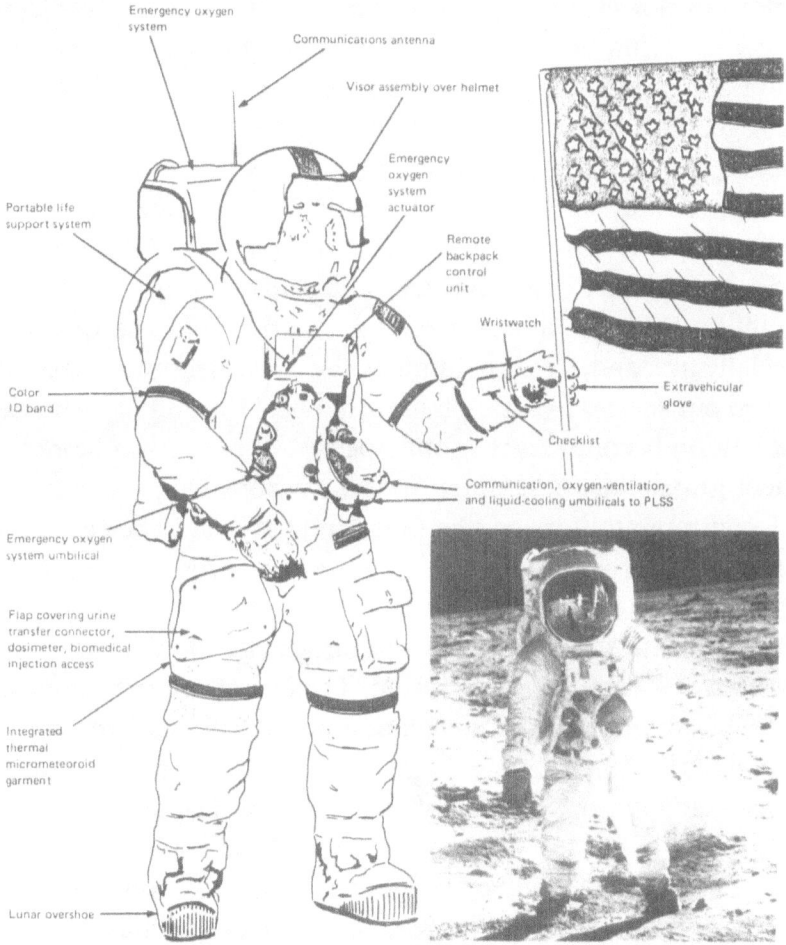

Figure 10.6 Apollo space suit. *Courtesy of NASA.*

DOES ·THIS CREATURE LOOK WELL ADAPTED TO HIS ZONE???

We have seen the difference between organic evolution and cultural evolution. Natural genetic evolution changes steadily over great time in erratic spurts during different periods, while human cultural evolution is exponential. This explains the overwhelming sense of rush in current technological development. It is changing exponentially. It is getting increasingly faster.

173

However, we humans are genetically stabilized. Actually, we employ our powers of cultural evolution to protect ourselves from the environmental pressures to change genetically. As a species, we are only cosmetically different from our ancestors of 40,000 years ago. And, there has been little pressure to adapt into another zone, or zones. Homo sapiens appeared with other hominids eons ago on the plains of Africa and adapted to the grasslands beyond the Rift Valley. Cooperative hunting in small groups of 20-30 individuals, deductive reasoning and strategic planning, the ability to see and scramble, the creation and use of tools, especially fire, and, overwhelmingly, the gift of self-awareness, all allowed our species to triumph in that zone. However, ages hence, the question becomes: is the same species, so perfectly adapted to ancient grasslands, best-suited to the environs of space culture? And, contrapuntally, need that grassland habitat of ancient Earth be recreated across the distant future just to protect our species from genetic change, just to stabilize and conserve our genome? Even an intuitive reaction tells us that genetic mobility must be chosen over stasis and conservation. The current cultural zone of 2,000 AD might well require adaptations, but simple projections reveal that the zones of space society will be manifold and require dramatic adaptations to each of those environments.

This is demonstrated by the mechanistic misconceptions put forth since early in the 20th century. The main notion has been to transport "up-there" and "out-there" by blaster, all the hardware and habitat necessary to establish the human environment of millennia past. From early on, the Moon was conceived as a trajectory for protected habitats as a first step into space. [4.] Space colonies were designed in the 1950s and various frameworks have been added since [4], but they all envision environments wherein sapiens can survive. While this past requirement is obvious to us now, we can also see it as limiting and even a dead end.

But within this major misconception we can find a way out, via its converse. It has seemed necessary to develop earth zones across the solar system because natural evolution has been accepted as a given. Serious scientific study has been devoted to the 'colonization" of Mars [6,] and it amply illustrates the oddity of the misconception. The whole process is compared to the colonization of the Americas. Manned habitats are deployed in every scheme. Sequential launchings deliver equipment, material and colonists in the vision of making the whole planet of Mars earth-like. Habitats protect sapiens from the hazards of cold, solar flares and cosmic rays [6,] Then, the long process begins to greenhouse the planet. This massive procedure is conceived over many many centuries by first releasing CFC gasses and warming the soil by various methods to unlock the CO_2 [6.] All of this is proposed to make a just-freezing earth zone. The whole scheme is to build a planet around sapiens rather than designing a hominid for the planet.

SAVING SPECIES The Alliance to Rescue Civilization differs from other so-called doomsday projects. It envisions a lunar base where, in the

The more examination that is given this central concept, the more obvious becomes its central flaw. The galaxy does not have an earth zone, and our grandest ambitions to occupy space will not

175

make it so. Artificial Evolution reverses this field and makes many a plan much more achievable.

We have examined the utilitarian and future generation arguments that favor Artificial Evolution. Here we have a functional argument for this new mode of evolution. The new adaptive zones of space present our species with a dilemma, an opportunity, and perhaps indeed, an inevitability. It would be as foolhardy to recreate a grassland zone across space as it would be to hold the human genome in stasis. Our species stares into a new environment and natural evolution is simply too slow and unpredictable to manage the adaptation. We must be able to function in the new zones of space or restrict ourselves to stasis. The bubble habitats conceived in the 20th century will be prisons confining us to creep about in physical and psychic servitude. A multiplicity of human species will free us to operate in and across the new zones of space.

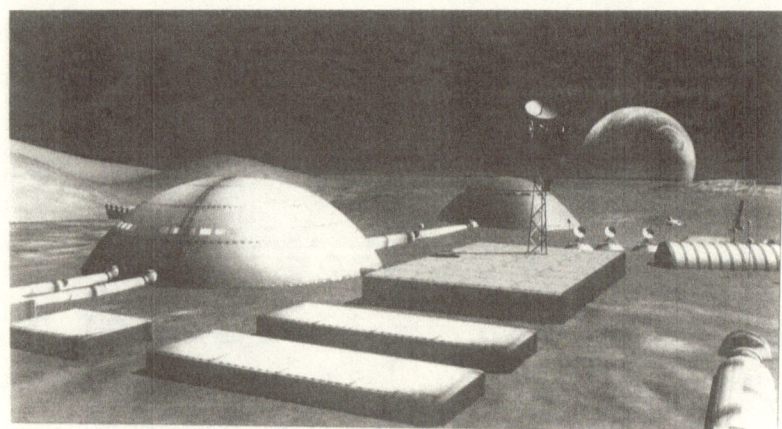

lunar colony could offer astronomers a clear view into deep space or the opportunity to mine rare minerals. The colony
could provide the flagship for a farflung space infrastructure. (M. Art)

THE CONCEPT OF BUILDING ENVIRONMENTS FOR OUR SPECIES ACROSS THE
SOLAR SYSTEM,WITH TERMS LIKE "SPACE COLONY" AND "OUTPOST", LEADS
TO DESIGN OF THIS TYPE. THIS IS CONFINMENT IN SPACE.

Figure 12.8 Unloading a habitat module just delivered to the lunar base. The flimsy looking crane can do the job easily because the module weighs only one-sixth its Earth weight. Other modules, delivered by other landers scattered around the scene, are already in place in trenches at the base in the background. The assembled habitat is then heaped over with lunar soil for protection and insulation. Earth hangs low in the sky. *Courtesy of Eagle Engineering; artist Pat Rawlings.*

```
THIS DEMONSTRATES THE LABOR AND CAPITAL INTENSIVE JOB  OF
CARRYING OUR EARTHLY ENVIRONMENT INTO SPACE.  JUST LOOK AT IT--
IS IT A"HABITAT MODULE" OR IS IT A PRISON CELL???

THIS DEPICTION ALONE INDICATES THE NEED FOR OUR SPECIES TO
EVOLVE ARTIFICALLY INTO THE SPACE ENVIRONMENT BEFORE US.
```

Naturally, this functional argument does not overcome all the physical confines of climactic restrictions. The surface of numerous planets and asteroids would never be hospitable to carbon-based life forms. Yet, synthesized adaptation to some challenging zones will be possible. More importantly, evolving our species to live within world cities, or tele-cities, around the sun is much more predictable. The imploded electronic culture that we have identified could be inhabited by beings much more attuned to those environments, ones much more in harmony to that new nature.

Thus, the pressure to evolve has been created by our cultural evolution. We stand at the verge of space society. We can now cultivate ourselves, quite literally, into its various adaptive zones and leave behind the misconceptions and bad planning of the 20th century.

While it seems that variations of our species would occur in our sensory and intellectual capacities, plain physical size must not be overlooked. Little beings will probably be better; they take up less space. The image of manned missions packed with 180 pound crew men crawling about cramped in cockpits becomes comical. 60

pound beings would be much more functional and space efficient. In the same sized space, there would be three times as many eyes, ears, hands and minds.

Figure 12.16 Artist's concept of a Mars base. Partially buried habitat is at the center, greenhouses to the right, and mine tunnels to the left rear. The vertical structure near the tunnel entrances is a well drilling rig. A "wagon train" vehicle made in segments climbs the hill in the foreground while a rocket lifts off at the right rear of the scene. Dish antennas on the mesa are for off-planet communications; the mast antenna is for communicating with roving exploration vehicles. Because Mars has an atmosphere, a very thin one, remotely piloted aircraft with very large wings can be used for transportation and exploration. In the lower right a geologist examines a newly discovered fossil. *Courtesy of Eagle Engineering; artist Pat Rawlings.*

ANOTHER EXAMPLE OF THE "COLONY" CONCEPT. DO THESE CREATURES LOOK WELL-SUITED TO THEIR ENVIRONMENT???

4 CREW MEMBERS AVERAGE WEIGHT IS 180 lbs. IF THE AVERAGE WERE 60 lbs THERE WOULD BE 12 CREW. THAT IS 3 TIMES MORE PEOPLE, AND 3 TIMES MORE HANDS, MIND POWER, COMPANIONSHIP, AND UTILITY. WITH ARTIFICAL EVOLUTION, HOMO sapiens novus CAN BE DESIGNED TO THIS SPECIFIC TASK. A 60 lbs HOMO sapien WITH SPECIAL ATTRIBUTUES WOULD AID TRANSPORTER MISSIONS CONSIDERABLY, AND WOULD BE MUCH NIMBLE UPON ARRIVAL ALMOST ANYWHERE IN THE SOLAR SYSTEM, INCLUDING SPACE STATION L-5.

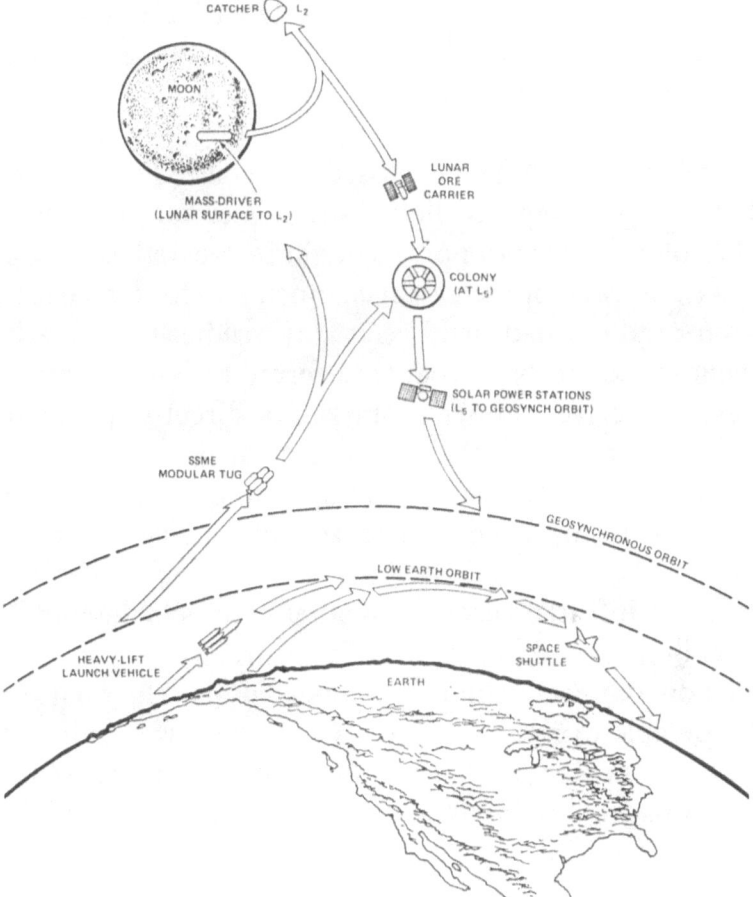

Figure 12.9 Cislunar infrastructure. (An SSME modular tug is a cargo vessel assembled from Space Shuttle main engines.) *Courtesy of NASA.*

THE HARD-WARE IS WELL PLANNED HERE, BUT IS THE SOFT-WARE???

HOMO sapiens novus CAN BE EVOLVED ARTIFICALLY TO UTILIZE THE ZONE ABOVE MUCH BETTER THAN THE HUMANS THAT EVOLVED UPON THE GRASSLANDS OF EARTH 40,000 TO 100,000 YEARS AGO. THIS ILLISTRATES THE UTILITARIAN ARGUMENT FOR MAXIMUM SUITABILITY AS OUR GREATEST GOOD, AND THE HIGHEST GOAL OF EDUCATED PEOPLE, WHICH COMES FROM J.S. MILL.

Clearly, new species would expand their sensory and mental capabilities. It is attractive to imagine more, just more – more eyes, more grey matter, more ears, etc. But, the argument here is more subtle. The basic structure of our senses and mind may be adapted to interface directly with the electronic circuitry of space culture. Indeed, our human nature may become as luminous as the new nature of space.

Yes, a larger cranium might be devised, with more synapses and even different pathways to allow more sequential processing. A number of new, sensory portals might be created, so beings could have two ears for the actual channel, two for the virtual, and two for radio broadcast. Even today, studies focus on the sonar gene-packs of bats that transferred to humans might allow us to receive radio transmissions directly. [1.] This might allow immediate transmission, or at least, reception, on broadcast channels. Such a development is an illustration of how space culture would condense and become unified.

Transplanting infrared vision from other species to humans is viewed as quite possible today [1]. In future, we may expand not only how our eye-brain links work, but also their nature. We may be able to add visual portals, as with the audio, so two eyes can be for actual and two for virtual, receiving electronic images directly into the cortex. Again, the species coalesces electronically.

Aging has been a central issue in bio-engineering and even in the old eugenics movement [2.]. While longer might seem better, in fact, numerous habitats might require shorter life spans so the genetic changes could occur more quickly. In any case, being able to vary longevity would be a key factor in space culture.

Other adaptations would be more critical, such as tolerances for atmospheric pressure. Some zones would be higher, some lower in pressure. Temperature tolerances would vary widely. Zones of higher and lower gravity would necessitate beings fitted to those requirements. And various atmospheres could call for a range of adapted hominids. Some might manage more or less nitrogen, oxygen and carbon dioxide.

The force of this argument is of necessity broad and speculative. We can only occupy space with new people, not old environments built across the solar system. Fairly soon we will have the knowledge and capacity to do it. But, the more particular becomes the prediction, the more unbelievable the position, so it is best to attend to the dynamic principle. The adaptations will come about not from fancy, but by specification. We will design beings for a particular environment just as we do a building or a machine for a task. It will be a practical made-to-measure sort of thing.

Functional necessity will prevail. To occupy the vast new niche of space we must evolve at a pace with our exponential cultural evolution, or remain stifled in stasis. Far from anything fanciful or futuristic, the bio-function described here seems inevitable.

So, space culture will be peopled with a new taxonomy of beings. They will be as varied as their environments. Yet, it seems probable that they will share a characteristic. They will be linked to the net of space, akin to its tissue, and a part of its fleshy membrane. The beings of space will be luminous, just as its architecture will be. Ray Kurzweil has projected the development of Artificial Intelligence over the next century, with "computers achieving the memory capacity and computing speed of the human brain by around the year 2020" [3.]. As computation speeds continue to grow, cubes rather than chips seem to offer many advantages. New methods are developing as well. Elongated carbon atoms, called nanontubes are under study, as are photons and DNA molecules themselves. But, moreover, digital computing may be replaced by quantum computing which would enhance capacities greatly [3,]

With Artificial Intelligence, we humans have created a thinking aid, a companion, and, in future, a co-habitant. Predictions for AI pose entities capable of thinking, creating and interacting much like, and very unlike, humans. Indeed, the human mind may be able to be loaded into AI so "we become software" [3, p. 150]. Although this may seem fanciful, the notion of transfer from carbon to a silicon-based experience can be extrapolated from the growth of the virtual. Virtual realities have not only invaded the actual and come to stand in for it; they increasingly pass for actual. As the presentation of sensory data grows in sophistication, so too will the verity of the virtual. Finally, virtual experience will be more compelling than the concrete because there will be more information. And, of course, silicon entities will process this experience more efficiently and directly than their creator, Homo sapiens, as they will be wired immediately into it.

Space culture will be woven of silicon entities of many types in an on-going development. However we name or characterize the way they live, they will be different from us, with both advantages and shortcomings. But certainly, the misconception of robot scouts preceding human settlers will be more than reversed, it will be nullified. The whole body of space will be Artificial Intelligence, its tissue AI, its glow, its beings, its very life, AI. This is why there will be no space. AI is the mode of its implosion.

Upon the verge of this new habitat, we have invented both Artificial Intelligence and Artificial Evolution. This conjunction begs a guess if not a hypothesis: both are necessary for surviving and exploiting the new niche of

182

space. The combination of AI and AE seems to be meant to happen. Cyborgs have been around for years and there are many examples today, such as jet planes, autos, aviation computer goggles, and even the average business office. So, the step to AI - Assists is only a technical problem soon to be overcome. Virtual images will be presented directly to the eyes; computational sequential thinking will be linked to the mind for immediate use. Homo sapiens will be adapted, and silicon devices will be fitted, to create this amalgam. Predicting what the exact devices will be is not useful; they will be much like those in use today, goggles, headsets, hoods, helmets, etc, but we will be rearranged for them. We will evolve new sense ports. The combination will be crucial to our suitability in space environments. Specific examples may just lead to apprehension and exaggerated reaction, but AI and AE together seems a certainty for our future. The symbiotic union will energize the lithe exercise of space culture.

The solar system is a stage set before us, and our species must choose wisely and deliberately how to people this play. Our new niche will be manifold and variable, so we will have to be as varied and mobile as the adaptive sets we create to live in. Our line of evolution will twist and turn in loops from our own hand, drawing backdrops that are jacquard, elegiac, or stark. The players will be a new Humanity, combined silicon-carbon, and the plots will be as unique as the protagonists. The fields of space will be alive with play, audience and actor as one event, interwoven and luminous. The show will be like fireflies of a summer night.

When we view the north-east corridor of the USA in satellite shots at night, we see a bright web of lights strung along the networks of human communiqué. This is alike the image of space culture to come. The brilliant lace will swirl out around the globe. The membrane will unwind about the sun in telecenters shining and abuzz. This space of no space will be virtual, imploded, symbiotic, transitive and luminous. Its nature will be ours. We will suit it.

References

1. <u>Discussions</u>. June 1998. Conference on Mammalian cloning. Washington, D.C.

2. <u>Ethics of Human Genetherapy</u>. 1997. Walters, L. and Palmer, J., Editors. Oxford University Press, NY

3. Kurzweil, R. 1999. <u>The Age of Spiritual Machines</u>. Viking-Penguin, NY, p. 3, 110, 15 FF

4. Lunan, D. 1983. <u>Man and the Planets</u>. Ashgrove, England. P.78 - 115

5. McLuhan, M. 1962. "The Electronic Age – The Age of Implosion" in "<u>Marshall McLuhan Essays</u>" 1997, Ed., Moss M., G&B Arts, 1997

6. Schmidt, S. and Zubrin, R. 1996. <u>Islands in the Sky</u>. Wiley and Sons, NY, p. 48 – 70, p. 125 – 145.

7. Virilio, P. 1991. <u>The Lost Dimension</u>, Semiotext, NY, p. 13 – 20.

CHAPTER 13

Conclusion

Once out of nature I shall never take
My bodily form from any natural thing,
But such a form as Grecian goldsmiths make
Of hammered gold and gold enameling
To keep a drowsy Emperor awake;
Or set upon a golden bough to sing
To Lords and Ladies of Byzantium
Of what is past, or passing, or to come.

Sailing to Byzantium,
W.B. Yeats.

It is a summer afternoon at the waterfront park of the Battery, New York City. Music floats across the lawns. Couples lie about bathing in the sun. The beat from radios and tape players intermix. Under a group of shade trees a young man plays upon a guitar. The whole pastoral scene is framed by grand post-modern architecture in a corner of old Manhattan where the street plan is centuries older than the uniform grid of later nineteenth century development. Yet the old wagon cart routes of the original port are echoed in the organic, fractured geometries of the new buildings. The natural paths of seventeenth century New Amsterdam are randomly revived and folded into the synthetic facades of the 21st century. And lo, beneath the hills of clad-steel, the melodies mingle.

The Hudson River churns past the cut granite levies, draining a quarter of the northeastern landmass as it has for millennia. Massive natural forces are channeled through a remarkable synthetic, man-made environment. Indeed, the park land is built upon fill. The post modern housing is founded upon the Hudson reclaimed, where now the musical melodies mingle on a sunny new millennium day.

Is the acoustic guitar in the park natural and, therefore, more authentic, more real? Or, is the digital, re-mastered disc music synthetic and thus more human, more real? Is the Hudson river water more real, or the bottled stuff that the sunbathers are drinking? Is the soil inland more real than the synthetic land-fill park? Judging seems inadequate, even though stating preferences is always satisfying. Just as it is somehow absurd to compare the ancient Hudson estuary to the man-made post-modern infrastructure, so too it is limiting to compare the natural acoustics to the disc synthesizers. The categories crumble as the melodies intermix and the glass clad facades mirror cloud, water, and shifting boughs, all built upon debris from our consumer society. Classes of things, zones of experience, and indeed, as we have seen, taxa of beings vaporize.

We find ourselves in the crux of categories confounded. The natural is crossed within the artificial. The elements of nature are so worked by the hand of man that the distinctions become prudish. In Artificial Evolution, the realm of culture and the realm of nature cross. The music is marvelous, the land-fill successful, and the bass spawn again in the Hudson.

Humans in the prehistoric world, and even nomadic societies of today, strictly separate the days in nature from those in culture, the mountain path from the city street, the dome of the sky from the flat ceiling confines of a bedroom. This ancient dichotomy represents a conflict in our development from nomadic and tribal society to settled, agricultural and city society. Of course, long before the rise of towns and settlements, this dichotomy did not arise, because in the distant past of our species, our technologies were rudimentary, like spears, knives, beads, and fire. But, today, even though various traditions in the East and West hold tight to the touchstone debate between the natural and cultural, in truth, they are crossed and confounded in a new entity. Ancient traditions based upon nature vs. culture dissipate in our present. Buddhist traditions that celebrate a closeness to nature must include the reflection of leaves from trees growing on landfill on glass facades as much as those reflections on the surface of a mountain pond. Adherents of secular humanism find as much delight and liberty from commuting across the Hudson as Christian monks do in contemplating its currents miles upstream at the Christian Study centers on the same Hudson River. In our world, the realms of nature and culture cross.

The natural and artificial are entwined, just as the pastoral of lower Manhattan reveals. But someday, we will not perceive the differences because we will live and breathe in a union of our culture and nature.

This distinction between man and nature, or at least, its perception, drives a great deal of debate about the ethics of biotechniques, as we have discussed. Our conclusions favored including our species in the natural realm. Clearly, Artificial Evolution is like other modifications of natural forces, and, consequently, its synthetic character is no more or less remarkable than a dam, nor is its effect more reprehensible. A new category arises and the natural slips inside the cultural just as our culture fits into nature. Indeed, the new laws of Artificial Evolution symbolize the crossing of categories, the binding of old distinctions. As a process, it is the combination of natural and synthetic forces. The fusion quells the debates based on the distinction. More than that, it points into a unified future. Our handiwork on the laws of evolution may prefigure similar modeling or re-structuring of other physical laws, allowing a human malleability of what we call "The Real".

Artificial Evolution exemplifies the process of transformation on the planet. Our species has shaped the most basic law, the life force, the advent of our coming, into our vehicle out beyond the globe.

We have described the new laws of Artificial Evolution, and placed them in natural philosophy, and speculated on their characteristics. We have seen how the new techniques differ both conceptually and culturally from the eugenics movements of the twentieth century. We have reviewed the format of ethical objections and examined the arguments that support this new mode of evolution, those arguments from Utilitarian analysis, from future generations, from value-progress in evolution, and from the advantage of multiple species. We have seen that individual choice and parental freedom will be the most likely method of implementation. Our study of climatic cycles and human populations has confirmed the necessity for Artificial Evolution. Our ability to endure new viruses and diseases gives further reason. But, survival, pure and simple, is the paramount biological factor. We must adapt and change to survive, so we might as well control it, rather than be its subject. Our humanity is experienced by real living people, so all those characteristics that we hold dear, and wish to insure, such as our ability to love, to appreciate life, to laugh, to create, to enjoy our families, will not continue, unless we adapt. The Gene Game does have a goal: it is the survival and adaptation of our Species.

Moreover, we have assessed throughout arguments or analytical points and they are just that; they are points of debate, arguments, pros and cons. We must not overlook the most important, indeed, outstanding truth of our current collective situation here on Earth: We have a magnificent opportunity. We can steer our genetic destiny and we must seize the chance. We have devised the way to manipulate a basic law in nature. We now have the tool to remake a critical aspect of reality. We can define a new our role in the world and become the modeler of laws, starting with genetic evolution and moving to other stricter regulations of physical and chemical reality. Hesitations and ethical questions must be honored and answered clearly. Indeed, all must be done in good measure, but the marvel of the opportunity is overwhelming. We really must seize this day.

In order for our culture to continue to develop, we must embrace the qualities of mobility and transitivity in our future world. The niche of space culture calls for the combination of AE-AI, so that our species can inhabit the territory we have already begun to explore. We can sense the lineaments of our future nature, the character of our future people and their role in our solar system and beyond.

We can see this new role for our species. Born into nature, like all the other species and surviving through our genetic capacity for culture, for thinking, for collective action, we are poised to step out of nature. By merging our cultural innovation with natural forces, we are changing our role in the world. Subjects to evolution, we are now to some degree the directors. We are remaking a central law of science and becoming a legislator of laws in the solar system. Our trangenetic and cloned creatures, as well as our own bio-procedures, demonstrate the wider process of Artificial Evolution, and it reveals a radical repositioning of our species. The initial phase of Artificial Evolution should be estimated for what it is. We are not only reforming the laws of evolution, we are coming into another station; we are other than our long genetic development. We are out of nature, as we are out of evolution and in a new relationship to the rest of nature. In the future, we may quite likely remake other laws of science and restructure the inanimate world to suit our new purposes. This new position creates complex relations, decisions, and ethical deliberations that individual bio-technologies only suggest. Yet, from the perspective of this new relationship, we can better assess the initiation of Artificial Evolution. Indeed, since the process provides us with a new role, we have another more profound reason to adopt it. The millions of years of hominid survival and development have brought us to this remarkable situation. The ability of "Lucy" has led to the creation of a Natural Law. Many of our limitations may be overcome and our future innovations may well bring about states of nature unimaginable and unknowable today. Artificial Evolution, then, is truly an adventure out of evolution, but it also creates a new voyager. The tool-maker of the prairies, the flint-clipper, the spear-maker, the stalker, the creature of city states and industrial automation has become the crafter of natural laws. The importance of the shift and monumentality of this new role can not be underestimated or understated. Once material to the laws of nature, we are becoming the modeler and legislator of those laws.

Note well that the player in The Gene Game is the legislator of laws. Our Species has become the legislator in this new world of our own making.

A New Role
Restructure The Laws Of Nature

This provides our species with a broad new role. It expands our perspective on the goals of science.

We can reconstitute aspects of reality by knowing and then remaking the principal laws of nature. Of course, the laws of physics and chemistry will probably be less malleable than those of biology. Nevertheless, our role has changed and expanded dramatically: we can know the truths of nature, and we can legislate new laws to reshape the real.

Our new form of evolution has opened an historic position for our species. We can now restructure some laws of nature.

But, why, and why at this time? Albert Einstein once asked, "How much choice did God have in constructing the Universe?" If the unified theory proposals of contemporary physics are correct, a creator or source would have had few design principles to work with, once the urge to make a Universe with star systems, DNA, and even inquisitive and self-aware minds was acted upon. But, our complete comprehension of "How" can never answer the question of "Why", for the impulse to act can not be explained by understanding its mechanics. However, as we answer the call to fit into the new niche of space and take up our new role as legislators, we could only be complimenting any possible process of an ultimate cosmology.

We concluded previously that there was no evidence for an ultimate cosmology driving the very human invention of Artificial Evolution. It is totally human. Indeed, there is not even a master plan or program on the human level. We can find no benevolent dictator, no master strategist, no evil genius. There is no program, no end in sight. This new mode of evolution is invented by us to address very human problems. We are not playing God, we are playing human.

In NATURAL THEOLOGY, 1802, the Reverend William Paley published the words that are so often quoted and stand referenced by Richard Dawkins at the front-piece of this book. The Reverend thought that we must conclude from the intricate marvels of the natural world, comparing them to a watch:

> "that the watch must have a maker, that there must have existed at some place or other, an artificer or artificers, who formed it for the purpose which we find it actually to answer; who comprehended its construction and designed its use."

Young Darwin thought and finally demonstrated otherwise. But, these words drive the historic natural evolution debates between the Evolutionists and Creationists, between the natural and divine source of the worldly realm. They explain the vehemence of Dawkins denial. We argue here that the invention of our new laws, and the establishment of our new role as Legislator, is distinct from theological ultimatums or any Universal Cosmologies. Humans are simply doing what they do best, whether divinely guided or not. We are winding the watch to a new time, to a new measure. We will pick its pace, quicken or slow it to our human purpose. Wonderfully and curiously, by holding this watchpiece so carefully and responsibly in our hands, we have brought into Universal Time the principles of the sacred. We have brought purpose and value into evolution. Of course, it is clearly human value and purpose, not Divine, be it divinely guided or not. This human legislative purpose will humanize our corner of the Gallaxy and then the larger Cosmos, as our ability to adjust the Laws grows and our responsibility matures. Curiously, then, whether the Watchmaker is real, known, ultimately knowable to us, or not, we are the Watchwinders. We have quite a time before us. Pity Darwin can not be along.

Artificial Evolution and Artificial Intelligence are inventions from expediency. Both have come about at the moment of space exploration, and, to some extent, because of it. Without reference to outside stratagems, these developments of the last century provide the methodology for our new role, the means to a future Galaxy, that are distinctly different from any technologies past. Our Species is reforming the basic situation and legislating laws, rather than only obeying them. Our specific techniques AE and AI will develop in fits and bounds, by trial and error and debate. The combinations will work in conscientious symbiotic evolution, in order to situate our descendants into those strange unearthly adaptive zones of our future. The Gene Game turns out to have serious, magnificent purpose: our survival, adaptation into, and creation of future space zones.

We forsee an anthropocentric future world, with our [species] at the top of the scale. While the procreative process is far from natural, it is certainly venerable, solemn, and sacred. Since our synthetic process thus brings us closer to those universals of mobility and change, we are more akin to nature and certainly the human enhanced nature of our future.

We need not know if a creative source, a God, condones, approves or envisions this process. Certainly, it does not deny or refute a universal God, but it is simply human. It is our process, the one we invented and can take full responsibility for, and one we can make humane. It is the doorway into our future, the path we will take to evolve out of evolution and into a value laden future of our own making. The tele-centers that we have characterized are the likely goal and reason for the process. We need to escape natural evolution and fuse the realms of culture and nature. We need to assume our new role. To survive the stresses upon our planet and to suit our new nature is reason enough. Truly, taking responsibility for our destiny and suiting our new environment is our greatest utility, our greatest common good, and a legacy of our long natural evolution. Our future will be done by us and for us in answer to the call from a vast new zone of space that lies before us.